AGAINST THE ODDS

AGAINST THE ODDS

Women Pioneers of Science

JOHN GRIBBIN
AND
MARY GRIBBIN

ICON

This edition published in the UK and USA in 2026 by
Icon Books Ltd, Omnibus Business Centre,
39–41 North Road, London N7 9DP
email: info@iconbooks.com
www.iconbooks.com

ISBN: 978-183773-132-9
ebook: 978-183773-131-2

Typeset by SJmagic DESIGN SERVICES, India

Printed and bound in Great Britain

Appointed GPSR EU Representative:
Easy Access System Europe Oü, 16879218
Address: Mustamäe tee 50, 10621, Tallinn, Estonia
Contact Details: gpsr.requests@easproject.com,
+358 40 500 3575

CONTENTS

	Acknowledgements	vii
	Foreword by *Jocelyn Bell Burnell*	ix
PREFACE	Overcoming Prejudice	xi
INTRODUCTION	Against the Odds	xiii
CHAPTER 1	Eunice Newton Foote (17 July 1819–30 September 1888) *Discovered the carbon dioxide greenhouse effect*	1
CHAPTER 2	Lise Meitner (7 November 1878–27 October 1968) *Explained the process of nuclear fission*	17
CHAPTER 3	Emmy Noether (23 March 1882–14 April 1935) *Found the mathematics that underpins particle physics*	39
CHAPTER 4	Inge Lehmann (1888–1993) *Discovered the inner core of the Earth*	55
CHAPTER 5	Irène Joliot-Curie (1897–1956) *Pioneered the development of artificial radioactivity*	67

CHAPTER 6	Cecilia Payne-Gaposchkin (1900–79) *Discovered what stars are made of*	85
CHAPTER 7	Barbara McClintock (16 June 1902–2 September 1992) *Discovered how genes turn on and off*	103
CHAPTER 8	Dorothy Crowfoot Hodgkin (12 May 1910–29 July 1994) *Used X-rays to determine the structure of vitamin B_{12}*	121
CHAPTER 9	Chien-Shiung Wu (31 May 1912–16 February 1997) *Discovered the law which allows matter to exist in the Universe today*	141
CHAPTER 10	Rosalind Franklin (25 July 1920–16 April 1958) *Used X-rays to determine the structure of viruses*	159
CHAPTER 11	Lucy Slater (5 January 1922–6 June 2008) *Pioneered the development of computer operating systems*	177
CHAPTER 12	Vera Rubin (23 July 1928–25 December 2016) *Discovered the widespread existence of dark matter in the Universe*	195
	Sources and Further Reading	215
	Index	221

ACKNOWLEDGEMENTS

Thanks to Paul Gribbin for insights into the workings of computer systems, and Ellie Gribbin for a critical reading of the early version(s) of the chapters. Lif Lund Jacobsen and Georgina Ferry gave permission for us to make extensive use of their studies of the work of, respectively, Inge Lehmann and Dorothy Crowfoot Hodgkin.

FOREWORD

From astrophysics and atmosphere to viruses and vitamins, this book holds the attention. It is both intriguing and infuriating – the women featured were so able, so innovative and so successful, and yet ignored, overlooked, diminished and patronised.

Twelve biographies cover almost two centuries from the birth of the earliest to the death of the most recent, and the penultimate chapter includes an excellent account (from personal knowledge) of the development of computer operating systems. The treatment is sympathetic, the science is clearly explained, but no punches are pulled – you can sense the authors' outrage at the treatment (or lack thereof) that these women experienced. The women are mostly Westerners, Chien-Shiung Wu being the exception, so this is also a reflection of how Western society's gender awareness has developed since the mid-1800s. (Answer – very slowly!)

What a long way science has come in those 200 years also! Technology, riding on the back of scientific discovery, has similarly boomed and enabled so much today. Imagine no laptop computers. Imagine not being able to contact relatives on the other side of the world. Imagine a pandemic like COVID without today's science and technology.

The accounts are good on context and full of intriguing snippets, which help humanise the women who might otherwise be lionised (a word with male connotations, interestingly) or set apart, above, on a pedestal. Above all, these are stories of intelligent observation, perseverance, staying power and, presumably, self-confidence.

Please read, enjoy and learn from these stories!
Jocelyn Bell Burnell
Oxford, May 2024

As a research student, Jocelyn Bell Burnell discovered pulsars, now known to be tiny, extremely dense stars on the edge of becoming black holes. She is a former Professor of Physics at the Open University, and a past President of the Royal Astronomical Society. In addition to many other honours, in 2007, she was appointed as Dame Commander of the British Empire.

PREFACE

Overcoming Prejudice

Even in the third decade of the twenty-first century, it is still harder for women to make a career in science than men, as many of our colleagues can testify. It may be easier than ever for them to get a foot on the ladder, but the higher up that ladder you go, the fewer women (in proportion to men) you will find. Two centuries ago, however, at the beginning of the nineteenth century, when science as we know it was just getting started, the situation was far worse. Then, the very notion of a female scientist would have been regarded as something of an oxymoron. Our aim in this book is to highlight the achievements of women who overcame the odds and achieved scientific success (although not always as much success as they deserved) in spite of male prejudice, as society changed over about 150 years, from the middle of the nineteenth century to the end of the twentieth century. This should be seen not only as a cautionary tale about the stifling effects of such prejudice, but as a celebration of those who achieved success against the odds – and maybe as an inspiration for the next generation. With that in mind, our last two chapters focus on two women who achieved success in science roughly at the same time, in the second half of the twentieth century, but did so through lifestyles which represent two extreme ways in which the odds could be overcome. Most women scientists today will find themselves somewhere in the middle ground between those extremes, but we hope through these examples to show just how big that middle ground is. If they can do it, you can do it.

INTRODUCTION

Against the Odds

There's a group of scientists who achieved some notable firsts over a span of roughly a century and a half. The identification of the carbon dioxide greenhouse effect; the theory of nuclear fission; the mathematics that underpins modern particle physics. The discovery of the Earth's inner core; the experimental evidence of nuclear fission; what stars are made of; how genes are turned on and off; the molecular structure of penicillin; experiments that established the 'standard model' of particle physics; the structure of DNA and viruses; the code that underpins computer operating systems; and the discovery of the dark matter that dominates the Universe. These scientists worked in different places, at different times, and mostly never met one another. So why do we call them a group? Because, of course, they had one thing in common. They were all women, working in a male-dominated environment.

How did they achieve such success, overcoming the hurdles that society placed before them?

Bente Rosenbeck, of the University of Copenhagen, has carried out a study of the lives of Danish female academics between 1875 and 1925, in which she identified four common features.* These apply so closely to the lives of the women discussed in this

* Har videnskaben køn? Kvinder i forskning, Museum Tusculanums Forlag, 2014.

book – whether Danish or not, and whatever decades they lived in – that it is worth spelling them out.

All the women came from families with either an academic background or money (or both); they all did better than the average man of their cohort in school or university; they all started on an academic path later than their male counterparts; and the great majority never married. In another study, this time of American scientists, Margaret Rossiter, of Cornell University, noted that at least up until 1940 there was a tendency for women to work in jobs that fitted the image society had of appropriate work for girls, but choose work close to the academic area they were interested in.* Many of them were also driven to prove that they were not just equal to, but better than male scientists, often working to the point of exhaustion or illness.

These are common themes of the careers of the scientists discussed in this book, even those working after 1940 in places other than Denmark and the USA. Some aspects feature more strongly in some lives than in others, but all of this could be an identikit description of the early career of Inge Lehmann, the subject of Chapter 4, who might be regarded as the archetypal example. She appears in Chapter 4, not Chapter 1, because we have chosen to arrange our subjects in the order of their birth, to highlight the (gradual!) improvement in opportunities for women scientists; but feel free to start with her if you wish.

* See bibliography.

CHAPTER 1

Eunice Newton Foote (17 July 1819–30 September 1888)

Discovered the carbon dioxide greenhouse effect

Most of the women who feature in this book achieved success after overcoming the prejudice against female scientists held by the scientific establishment at large and by most male scientists. Their stories show how these barriers were gradually eroded during the twentieth century. But our first pioneer is something of an anomaly, a woman who made an important scientific discovery in the nineteenth century, although its importance was not appreciated until quite recently. Her success actually highlights the problems of women in science at the time. Far from being a professional physicist working in an academic environment, she was a wealthy amateur who, with the support of her husband, was able to set up her own laboratory and work without any need to satisfy the male hierarchy. She was, though, well aware of her privileged position, and campaigned vigorously not just for women's rights but for other causes, such as the abolition of slaves. This may have overshadowed her scientific work – which may also have been overlooked partly because of the unfortunate coincidence that she and her husband shared the same initial on their forenames.

Eunice Newton (as she then was) was born on 17 July 1819, in the small town of Goshen, Connecticut. Her father claimed to be a distant relative of Isaac Newton, who had died in 1727, placing her much closer in time to him than to us. The year after Eunice was born, the family moved to Ontario County, in western New York State, where her father owned a farm. She had five brothers and six sisters, one of whom took over the farm after their father died in 1835. We know what she looked like from the description on her 1862 passport application, when she was in her early forties: 'just under 5 ft 2 in [1.57 m] tall, with blue-gray eyes, a rather large mouth, with an oval face, a sallow complexion, and dark brown hair'.

The region where Eunice grew up was a hotbed of social and political activism, involving campaigns supporting the abolition of slavery, women's rights and temperance. Reflecting these attitudes, she was educated at a pioneering school called the Troy Female Seminary, founded by the feminist Emma Willard, who believed that women could handle mathematics and science as well as men; to this end, her students attended science classes, including hands-on laboratory work, at the nearby Rensselaer School. The seminary was the first school in the USA to offer higher education for women, including mathematics, science, history, geography and philosophy. Eunice had a grounding in all these subjects, plus French, by the time she graduated in 1838. But there was no career path in science for even a well-educated and scientifically literate woman in those days, and it is no surprise that she married (a lawyer, Elisha Foote), on 12 August 1841, when she was just 22 and he was 32. The marriage produced two daughters, Mary in 1842 and Augusta in 1844.* But

* Like her mother, Mary became an active campaigner for women's and other rights. She married John B. Henderson, the senator from Missouri who introduced the amendment to the Constitution that abolished slavery. Henderson was also one of the seven Republicans who voted against the impeachment of President Andrew Johnson. Mary died in July 1931, a week short of her 89th birthday.

Eunice was no conventional housewife, and her partnership with Elisha produced a lot more than children. She was an active women's rights campaigner and was on the editorial board of the 1848 Seneca Falls Convention, the first conference on women's rights, organised by prominent suffragist Elizabeth Candy Stanton, a friend and neighbour of the Footes. As a member of the editorial committee for the convention, Eunice – and her husband – signed the convention's Declaration of Sentiments, which demanded social and legal rights equal to those of men, including the right to vote.

Elisha Foote specialised in patent law, and made several inventions of his own. In the 1850s, the family were still living in Seneca County, New York, where the couple were affluent enough to have a scientific laboratory built in their house. There, separately and together, they carried out experiments which were a cut above the usual dabbling of scientifically minded gentlemen (and women) amateurs. One pair of these experiments was reported in two papers, one by Elisha and one by Eunice, read to the tenth annual meeting of the American Association for the Advancement of Science (AAAS), held on 23 August 1856 in Albany, New York. The two papers were published later that year in the *American Journal of Science and Arts*. Eunice's represents the first known publication on experimental physics by an American woman in a scientific journal (there had been earlier publications by women on observational astronomy), and is the basis for the claim that she was the 'discoverer' of the global-warming greenhouse effect.

The two papers are best considered as a pair, since clearly the couple were working on a joint project to study the way the Earth is warmed by the Sun. Elisha reported experiments on the temperatures recorded by thermometers exposed to the sun or shade, inside or outside a room, at various air temperatures. Eunice investigated the effect of the Sun's rays on different gases, sealed inside two glass containers. Each glass cylinder, four inches in diameter and thirty inches long, contained two thermometers to ensure that the temperature readings were reliable, and the two cylinders could be connected together via an air pump, which she used to empty one cylinder into the other, where it was 'condensed', in her terminology. After allowing both cylinders to reach the same temperature in the shade, she then placed them in the Sun and measured the temperature in each cylinder once it had stabilised. She repeated the experiment with wet air and dry air, and with hydrogen or carbon dioxide (known to her as carbonic acid) replacing the air. The results were clearcut. Containers filled with moist air or with carbon dioxide warmed up more than containers filled with dry air, and carbonic acid trapped the most heat, with the temperature in the tube reaching 125°F (52°C).

The fact that containers made of glass could trap solar heat was not in itself a surprise in the 1850s. After all, a room with closed glass windows gets hotter than the air outside when the Sun's rays pass through the glass. This phenomenon had been studied by the eighteenth-century naturalist Horace de Saussure, who was a native of Geneva and was particularly interested in the meteorology and geology of the Alps. He made a glass box, sometimes called a solar oven, and in 1767 measured the temperature inside when it was exposed to sunlight on the plains of Cournier and on the top of Mount Crammont, 4,852 feet above the plains. In both places the temperature inside the 'oven' reached (in modern units) 110°C, although the temperature of the outside air on the mountain top was 19°C lower than that on the plains. This highlighted a widely discussed puzzle of the time – why are valleys warmer than mountains even though mountains are closer to the Sun?

The next step, building on the work of De Saussure, came from the French mathematician and physicist Joseph Fourier in the 1820s. Fourier studied the way heat is transferred between objects, and this enabled him to calculate how hot an airless globe at the same distance from the Sun as the Earth would be if it was solely heated by solar radiation. The temperature he calculated is much lower than the actual average temperature of the Earth – but, of course, the Earth is not airless. Bringing things up to date, we now know that, averaging over day and night and all latitudes, the average temperature of the Moon, which is airless, is about -18°C, while the average temperature of the Earth is about 15°C. Fourier considered several possible reasons for the warmth of the Earth, and favoured the (incorrect) idea that our planet is heated by radiation from space. But he also mentioned De Saussure's work, and suggested the possibility that the atmosphere of the Earth acts like a blanket around the planet, trapping heat that would otherwise escape into space. In 1824 he came very close to identifying the mechanism by which this works. He wrote 'the temperature can be augmented by the interposition of the atmosphere, because heat in the state of light finds less resistance in penetrating the air, than in repassing into the air when converted into non-luminous heat'.* This is now widely regarded as the first hint in a scientific publication of what is today called the terrestrial greenhouse effect, although Fourier did not use that term and, crucially, he did not know which gases in the atmosphere 'resist' the passage of the outgoing radiation from the warm surface of the Earth.

Eunice Foote's improvement on this idea was to identify two of the gases which do indeed trap heat – carbon dioxide and water vapour – and then to link this discovery with the idea that the changing concentration of these gases in the air could contribute to climate change. Although she did not give any references in her paper, it is quite likely that she knew of Fourier's work. She said in the introduction to the 1856 paper: 'My investigations have had

* See http://fourier1824.geologist-1011.mobi/

for their object to determine the different circumstances that affect the thermal action of the rays of light that proceed from the sun.' And after explaining the experimental setup she described what she had observed:

> The highest effect of the sun's rays I have found to be in carbonic acid gas …
>
> An atmosphere of that gas would give to our earth a high temperature; and if as some suppose, at one period of its history the air had mixed with it a larger proportion than at present, an increased temperature from its own actions as well as from increased weight would have necessarily resulted.

This is the basis of the suggestion that Foote 'discovered' the terrestrial greenhouse effect; but the recognition of how the effect works came just three years later, when the Irish physicist John Tyndall, who was unaware of Foote's work, carried out a series of experiments at the Royal Institution in London on the absorption of what was then called 'radiant heat' (Fourier's 'non-luminous heat', now known as infrared radiation) by gases.

In the early nineteenth century, the astronomer William Herschel had discovered that radiation invisible to human eyes, with longer wavelength than red light (hence the name 'infra-red'), could warm a thermometer. In the 1850s, John Tyndall was an established scientist with a senior position at the Royal Institution, one of the leading scientific research centres of the time, and access to the best experimental apparatus of the day. He was familiar with the work of Herschel, De Saussure and Fourier, and in the spring of 1859 he began a series of experiments to measure the way radiant energy is absorbed by different gases. As a source of infrared radiation, he used a small copper tank containing boiling water at 100°C, with the radiation passing

Eunice Newton Foote
Public domain, via Wikimedia Commons

through a long tube which could be filled with different kinds of gas. He could measure both the temperature of the gas and other properties using sensitive detectors. On 26 May that year he wrote to the Royal Society announcing his programme of research 'on the transmission of radiant heat through gaseous bodies', pointing out that 'We know nothing of the effect even of air upon heat radiated from terrestrial sources', and: 'With regard to the action of other gases upon heat, we are not, so far as I am aware, possessed of a single experiment.'

As well as making it clear that he was unaware of Foote's work, this hints at a crucial difference between his work and hers. She assumed that the air in her cylinders had been warmed by the Sun's rays on the way in. We now know that infrared radiation is blocked by glass, so there was little or no direct heating of this kind in her experiments. But when the walls of such a container are warmed, by absorbing solar energy, they emit infrared radiation, and this is what is absorbed by gases such as carbon dioxide and water vapour, raising the temperature of the gas in the tube. Tyndall realised this. On 10 June he demonstrated his first experiments on radiant heat to the Royal Society, and said that although solar heat can get through the atmosphere to warm the surface of the Earth:

> when the heat is absorbed by the planet, it is so changed in quality that the rays emanating from the planet cannot get with the same freedom back into space. Thus the atmosphere admits of the entrance of solar heat; but checks its exit, and the result is a tendency to accumulate heat at the surface of the planet.

This, in a nutshell, is what we now call the greenhouse effect. After further research, measuring the ability of many gases, including carbon dioxide, to absorb infrared radiation, Tyndall summed up his work in a major lecture given to the Royal Society in 1861:*

> These experiments were commenced in the spring of 1859, and continued without intermission for seven weeks. The course of the inquiry during this whole period was an incessant struggle with experimental difficulties. Approximate results were easily obtainable, but I aimed at exact measurements … With this apparatus I determined approximately the absorption of nine gases and twenty vapours during the summer of 1859.

He also mentioned the possibility that a change in the concentration of gas such as water vapour in the air might cause 'mutations' in the climate.

So how come Tyndall knew nothing of Eunice Foote's work, which was largely forgotten for more than a hundred years? There are several reasons, including the fact that she lived in America, which was very much a scientific backwater in the second half of the nineteenth century. To put this in context, at about the same time that Tyndall began his research on the topic, in 1859, the German scientist Gustav Magnus was carrying out similar experiments in Berlin. But neither of them knew about the other one's work, until Magnus learned of Tyndall's first paper nearly two years after it had been published and wrote to him. It is hardly surprising that neither of them knew anything about the work of an amateur female scientist, published in an obscure American journal. But at least Foote's work did receive some recognition in that backwater.

* J. Tyndall, 'The Bakerian Lecture: on the absorption and radiation of heat by gases and vapours, and on the physical connexion of radiation, absorption, and conduction', *Philosophical Transactions of the Royal Society*, volume 151, pages 28–29 (1861).

Although Elisha Foote read his own paper to the AAAS meeting in 1856, Eunice's paper was presented on her behalf by Joseph Henry, the first director of the Smithsonian Institution. There was no rule forbidding women from speaking at the meetings of the Society, but it would have been unusual, and we do not know whether she was actually present at the meeting. Having Henry read her paper may have indicated its merit to the audience, but, unfortunately, Henry prefaced his presentation with comments drawing attention to the fact that the author was a woman, rather than concentrating on the scientific results. His remarks were recorded by David Wells, the editor of a journal called *The Annual of Scientific Discovery*, who reported in his 1857 volume (looking back to 1856):*

> Prof. Henry then read a paper by Mrs. Eunice Foote, prefacing it with a few words, to the effect that science was of no country and of no sex. The sphere of woman embraces not only the beautiful and the useful, but the true. Mrs. Foote had determined, first, that the action of the rays increases with the density of the air. She has taken two glass cylinders of the same size, containing thermometers. Into one the air was condensed, and from the other air was exhausted. When they were of the same temperature the cylinders were placed side by side in the sun, and the thermometers in the condensed air rose more than twenty degrees higher than those in the rarified air. This effect of rarefaction must contribute to produce the feebleness of heating power in the sun's rays on the summits of lofty mountains. Secondly, the effect of the sun's rays is greater in moist than in dry air. In one cylinder the air was saturated with moisture, in the other dried with chloride of lime;

* *Annual of scientific discovery: or, year-book of facts in science and art, for 1857*, Gould and Lincoln, Boston, 1857.

> both were placed in the sun, and a difference of about twelve degrees was observed. This high temperature of sunshine in moist air is frequently noticed; for instance, in the intervals between summer showers. The isothermal lines on the earth's surface are doubtless affected by the moisture of the air giving power to the sun, as well as by the temperature of the ocean yielding the moisture. Thirdly, a high effect of the sun's rays is produced in carbonic acid gas. One receiver being filled with carbonic acid, the other with common air, the temperature of the gas in the sun was raised twenty degrees above that of the air. The receiver containing the gas became very sensibly hotter than the other, and was much longer in cooling. An atmosphere of that gas would give to our earth a much higher temperature; and if there once was, as some suppose, a larger proportion of that gas in the air, an increased temperature must have accompanied it, both from the nature of the gas and the increased density of the atmosphere. Mrs. Foote had also tried the heating effect of the sun's rays on hydrogen and oxygen, and found the former to be less, the latter more, susceptible to the heating action of sunlight.

The September 1856 issue of *Scientific American* was also impressed that a woman could be a scientist. An article on 'Scientific Ladies – Experiments with Condensed Gases' focused on her work, commenting 'This we are happy to say has been done by a lady' who 'was deeply acquainted with almost every branch of physical science' and 'the experiments of Mrs. Foot [sic] afford abundant evidence of the ability of woman to investigate any subject with originality and precision'. The work was also mentioned in the *New-York Daily Tribune* and in both a German review of the year's science and in the *Edinburgh New Philosophical Journal*, although neither of them included her conclusions about carbon dioxide and climate. The Edinburgh journal also gives us a hint of why Eunice's work received less attention than it might have. It refers to two papers, one by Elisha Foote and one by 'Mrs Elisha Foote'. And although it only gives the title of Elisha's paper ('On the Heat in the Sun's Rays') it mostly describes her work! The report is so short that it is worth presenting in full:

> On the Heat of the Sun's Rays. By Elisha Foote. On the Heat of the Sun's Rays. By Mrs Elisha Foote. Read by Professor Henry: These papers described experiments from which it was inferred, that the heating power of the sun's rays varies with the temperature of the place into which the rays fall, that the temperature of air is raised by sunshine passing through it, that in the same condition rarefied air is less heated than that which is condensed, moist air more than dry air, carbonic acid gas more than atmospheric air, and oxygen more than hydrogen gas.

The reader might naturally assume that, at best, she was Elisha's assistant.

Whatever the reasons for Foote's 1856 paper failing to make much impact, the following year she made a breakthrough of greater significance in the history of women in science, showing that she was no one-hit wonder. After her experiments on the thermal properties of gases, Foote turned her attention to their electrical properties. She used an air pump to vary the pressure of gases in a glass tube about two feet long and three inches in diameter (not the same as the tubes used in her earlier experiments) and measured the ability of different gases, at different pressures, to hold a charge of static electricity. The gases she tested were dry and damp air, oxygen, hydrogen and carbon dioxide, and the electric charge was measured using a device called an electrometer, where two thin gold leaves were pushed apart by electric repulsion. Her paper, titled 'On a New Source of Electrical Excitation', was presented on her behalf by Henry at the 1857 meeting of the AAAS, on 14 August that year, in Montreal. Unlike her earlier paper, this one was published in the more prestigious *Proceedings of the American Association for the Advancement of Science*, in November 1857. It was the first publication on physics by a woman in that journal. Also unlike the 1856 paper, this time Foote included references to earlier work by other scientists, including the Frenchmen Joseph Gay-Lussac (who discovered that water is made of two parts hydrogen and one part oxygen) and Jean-Baptiste Biot (who proved that meteorites are rocks from space), showing that she had read their papers and had much more than a dilettante interest in the subject.

Foote's results showed that the amount of static electricity generated in her tube was related to the moisture of the air, and that she could change the moisture content by expanding or compressing the air. The opening line of her paper reads: 'I have ascertained that the compression or the expansion of atmospheric air produces an electrical excitation', and she elaborates that:

> My experiments with this apparatus have extended over about eight months, and I have found the action to bear a strong analogy to that of the electrical machine. In damp or warm weather little or no effect would be produced, whilst at other times, particularly in clear cold weather, the action would be so strong as to diverge the leaves of the electrometer to their utmost extent. In warm weather when no action would be produced, I have attained the result by cooling the air artificially. A sudden expansion or contraction always increases the effect.

She also points out the importance of this in understanding the behaviour of the atmosphere of the Earth:

> Particularly should this be observed in the dry cold regions of our atmosphere above the effects of moisture and vapors; and it

was established by the experiments of Becquerel as well as those of Gay Lussac and Biot that the electricity of the atmosphere increases in strength with the altitude.

This paper, published in a more prestigious journal, attracted greater attention in Europe and was reprinted in the *Philosophical Magazine* in 1858.* Tyndall must surely have seen it, but there is nothing in the paper to connect it with her earlier paper. This, however, is the last we hear of Eunice Foote the scientist, until, thanks to the second paper, her name begins to be noticed by historians in the last quarter of the twentieth century, and then the significance of her paper on atmospheric warming becomes recognised in the first quarter of the twenty-first century. We don't know why she stopped doing scientific experiments – indeed, we don't know why, or when, she started doing them. But there may be a clue in the fact that sometime in the late 1850s, Elisha and his family moved from Seneca Falls to Saratoga Springs, where he developed his practice in patent law. Both Elisha and Eunice patented several inventions over the years that followed, so her attention may have shifted from the abstract to the practical.

In 1865, Elisha was appointed to a post at the US Patent and Trademark Office, and the family moved on to Washington, DC, where he rose through the ranks to serve as commissioner of patents from July 1868 until 1869. Their daughter Mary married in 1868, and her sister, Augusta, the following year. After several career-related moves, with the children having long since left home, the couple were living in New York City by 1881. Elisha died in 1883, and Eunice at Lenox, Massachusetts, on 30 September 1888.

* Where her name was given as 'Mrs Elisha Foote'! *Philosophical Magazine*, volume 15, pages 239–40, 1858.

As the first time an American woman's work on physics was published in a major journal, *Proceedings of the American Association for the Advancement of Science*, her 1857 paper, although her second published paper, is the more significant in terms of the impact of women breaking down barriers in science, even though its subject matter is of less scientific importance today. In the whole of the nineteenth century, only sixteen physics papers were published by American women; and the only two published before 1889 were those by Eunice Foote. In 2022, the American Geophysical Union established The Eunice Newton Foote Medal for Earth-Life Science, awarded to recognise important scientific research on the links between Earth- and life- sciences.

The important message to take away from all this is not who was the first person to suggest the existence of a terrestrial greenhouse effect. As is often the case in science, several people were working along similar lines at about the same time. The significance of Eunice Foote's story is that it is an early example of the truth that, given the opportunity, women can do science just as well as men. The *Scientific American* article made the same point: 'Owing to the nature of women's duties, few of them have had the leisure or the opportunities to pursue science experimentally, but those of them who have had the taste and the opportunity to do so, have shown as much power and ability to investigate and observe correctly as men.'

CHAPTER 2

Lise Meitner (7 November 1878–27 October 1968)

Explained the process of nuclear fission

Lise Meitner was born at just the right time to push open the door preventing women from becoming professional research scientists. But this accident of birth also meant that for complicated reasons not entirely related to her gender, her greatest contribution to physics did not receive due recognition when it was carried out, and her place in history is still often overlooked in favour of a male colleague. There are many examples of the Nobel Committee failing to recognise work worthy of their notice. Even Albert Einstein's work on relativity was overlooked, although he did receive the Prize for another contribution. But Meitner's is perhaps the clearest example* of someone unjustly deprived of the accolades due to her – not least because the male colleague did get the Prize for their joint work, in which if anything she was the team leader.

Meitner was born in Vienna in November 1878. She came from an essentially secular Jewish family, and her birth was noted in the

* Only 'perhaps' because of the bizarre failure of the Committee to recognise the work of Fred Hoyle, but that is another story.

records of their community as the 17th of that month, but in all other documents the date is given as 7 November, which was the date on which she always celebrated her birthday. Her paternal ancestors came from a village called Meiethein, in what is now the Czech Republic, from which the surname Meitner derived. Her mother's family descended from Jews who had fled Russia to avoid persecution. Philipp and Hedwig Meitner, who married in 1873, had eight children, Lise (initially known as Elise) being the third, and the youngest, Walter, arriving in 1891.

At that time, the Austrian laws forbade women from attending university, or even from receiving any education beyond the age of fourteen. Just before she reached that age, in the summer of 1892, Lise left school with no prospects except marriage or becoming a teacher in a subject, such as French, which did not require an advanced education. For five years, Lise was stuck at home, helping with domestic duties. Then, in 1897, under pressure from feminist activists, the government relented. It was decreed that women would be allowed to enter universities to study, provided they passed the Matura, the standard school-leaving examination, whether or not they had attended the Gymnasium (High School). After finishing a teacher-training course as a backup, Lise began cramming for the Matura with two other girls, supervised by a physicist from the University of Vienna. In July 1901 she was one of fourteen women who took the exam; only four passed – her cohort of three and Henriette Boltzmann, the daughter of a famous physicist. In October, Meitner began studying science at the University of Vienna. The quality of the physics teaching there was first class. In the first year, she was taught by Franz Exner, a pioneering researcher in the medical applications of X-rays, and from her second year onwards she was under the influence of Henriette Boltzmann's father, Ludwig, who had explained the laws of thermodynamics in terms of the behaviour of atoms and molecules, and who was an inspirational teacher. She became hooked on physics for its own sake, whether or not there was any prospect of a career in science.

In Vienna at that time the university operated slightly differently from the way universities operate today. After she finished her coursework in the summer of 1905, Meitner was able to spend a few months working on an experimental research project, under the supervision of Franz Exner, which led to the award of a doctorate in February 1906; this would be about equivalent to an MPhil today.* She was only the second woman to be awarded this degree in physics from the University of Vienna, but no female physicist had made the next step up the academic ladder to become a so-called assistant. For the rest of 1906, Meitner stayed on at the university carrying out some experiments in optics and, significantly for her future, with the newly identified alpha and beta forms of radioactivity (measuring how the beams were scattered by metal foil), teaching at a school in the daytime and carrying out research on radioactivity in the evenings under the supervision of Stefan Meyer. But with no prospects in Vienna, the next year she asked her parents (who were still supporting her financially) for permission to try her luck for a few terms in Berlin, which was the centre of the German-speaking scientific world. She arrived there in September 1907, shortly before her twenty-ninth birthday, and would stay for three decades.

This was the year before women would be admitted as students to Prussian universities, although they could 'audit' lectures without taking examinations or being awarded degrees, and even as the holder of a doctorate Meitner still had to ask the permission, only grudgingly given, of Max Planck, the founding father of quantum

* This was the year that a certain Erwin Schrödinger began his own studies at the University of Vienna.

physics, to attend his lectures. She also wanted to broaden her experience of experimental work, so she got permission from the head of the physics laboratory, Heinrich Rubens, to work there. He introduced her to another young researcher, Otto Hahn, with the suggestion that they collaborate on a project. It was a marriage made in scientific heaven, although it would end in tears.

Hahn was four months younger than Meitner, but he had more experience, as he had followed a conventional path through the male-oriented academic system of the time. He studied chemistry at the University of Marburg, worked on radioactivity first with William Ramsay in London then with Ernest Rutherford in Canada, and came back to Europe in 1906 with an appointment as assistant in Berlin. A year later he was promoted to the rank of privatdozent, the most junior of lecturers.

There was one problem with the proposed collaboration with Meitner. The chemistry laboratory was completely barred to women. In a striking echo of the way Marie Curie had been treated a few years earlier in Paris, Meitner was allowed to work in a basement which had its own access from outside. If she needed the toilet, she had to visit a restaurant nearby. When the law changed in 1908 she was allowed upstairs, and the department even installed a toilet for 'ladies'.* But there was still considerable hostility to women, as she recalled in her memoir *Looking Back*: sometimes when she and Hahn bumped into a colleague together, the colleague would pointedly say, 'Guten tag, Herr Hahn,' while ignoring her. The physicists, she noted, were generally much more friendly.

The collaboration was an immediate success. The study of radioactivity straddles the boundary between physics and chemistry, with Meitner making measurements of radioactivity and

* Toilets, or the lack of them, will be a recurring theme in this book.

analysing the results as a theorist, while Hahn carried out the chemical processes that separated radioactive elements from mixtures of material. In August 1908, Meitner went home for a holiday and asked her parents to continue funding her for another year. She also, for reasons she never explained, formally left the Jewish community and on 29 September she was baptised as a Protestant.

Back in Berlin, in December the team, becoming well known for their publications on radioactivity, received a visit from Rutherford, on his way home from receiving the Nobel Prize. On being introduced to Meitner, he exclaimed: 'Oh, I thought you were a man!'* It is not clear whether this was intended as a compliment.

With the research going well, Meitner persuaded her parents to continue her modest allowance, which she supplemented by translating articles from English into German and by writing for a popular scientific magazine, where her articles were signed 'L. Meitner'. Impressed by her output, the editor of an encyclopaedia wrote to ask 'Dr L. Meitner' to contribute an article, but on learning her first name responded: '[I] would not think of printing an article written by a woman!'†

Even with this additional income, there was little to spare for luxuries. Meitner lived alone in a succession of rented single rooms, considered a newspaper a luxury and could just about afford the cheapest seats at concerts. Her holidays were spent hiking in the mountains. At work, she began to give lectures, and attended a conference in Saltzburg in September 1909 where she met Albert Einstein, who was, to the month, the same age as Hahn.

In the years up to 1912, Hahn and Meitner published a string of papers, and their work was recognised when the new Kaiser

* *Looking Back*, see Further Reading.

† Hahn, *A Scientific Autobiography*.

Wilhelm Institute (KWI) for Chemistry was founded. Hahn was offered a post there as professor, in charge of the radioactivity section, with a salary of 5,000 marks. And Meitner would be welcome to work alongside him – as an unpaid 'guest'. This situation was even more precarious because her father had died in 1910, jeopardising her allowance from him. Then, out of the blue, late in 1912, as the new KWI was opening its doors, Planck appointed Meitner as the first female assistant in Prussia. For the first time in her 34 years, she would be paid (a little – 1,500 marks per year) to do physics. The following year, she became an official scientific associate in the radioactivity section, which became known as the Hahn–Meitner section. Then, everything changed.

The year 1914 started well. Around the time Meitner moved to Berlin, Hahn had discovered a radioactive element then called mesothorium, which was later identified as an isotope of radium. Meitner helped him devise a technique to purify samples of this material, which turned out to have practical applications in medicine. The process was developed on a commercial scale by a Berlin company, and in 1914 Hahn received 66,000 marks in royalties, and passed on 10 per cent to Meitner. This was followed by 40,000 marks, shared in the same way, in 1915. But the arrangement was ended by the war.

When the war began, Hahn and Meitner had just started a project in their new laboratory, to find the 'precursor' of the radioactive element actinium. Actinium is produced by the radioactive decay of uranium, but not in a single step; the evidence showed that a short-lived intermediate product had to be produced from the uranium, with this precursor then almost immediately decaying into actinium. The search for the precursor was exactly the kind of job for which the combined talents of the Hahn–Meitner team were suited. And Meitner had at last begun to receive the financial rewards she merited; in August 1914 the KWI, prompted by learning that she had been offered a job in Prague and being anxious not to lose her, increased her salary to 3,000 marks. But a few weeks later, Hahn was in the army* and Meitner, gripped by the patriotic fervour which initially infected all of the belligerents in the First World War, decided to train as a medical X-ray nurse/technician, while keeping work at the lab ticking over.

In 1915 and 1916, Meitner worked in her new capacity with the Austrian army on the Russian front, where as well as the X-ray work, she carried out conventional medical duties, including assisting at operations and nursing the dying. She was then transferred south to the Italian front before returning to Berlin in October 1916. In January 1917 she was appointed to head her own section at the KWI, with her salary increased to 4,000 marks – still 1,000 less than Hahn's salary, but his included a marriage allowance, and she never married, nor showed any inclination to do so. Over the following months she completed the actinium work, and the discovery of a 'new' element that became called protactinium† was announced in a joint paper by Hahn and Meitner, published in March 1918. After this, Einstein, who was then based in Berlin, took to referring to her as 'our Marie Curie'. But Germany was descending into chaos and hunger as its armies were defeated and extremists from left and right fought in the streets. Out of that chaos a new beginning eventually emerged, and one of the first signs of it for Meitner was that in the summer of 1919 she was formally made a professor – the first woman in a German university to receive that title.

Nevertheless, that same year the Association of German Chemists awarded their top honour, the Emil Fischer Medal, to

* In 1915 he was recalled from the Belgian front to join Fritz Haber's team developing poison gases, and later saw their effectiveness against the Russians. He was, of course, just obeying orders.

† Originally protoactinium, but the name was soon shortened.

Otto Hahn alone, citing among other things 'his' discovery of protactinium, and gave Meitner a copy of the medal in a half-hearted attempt to recognise her contribution. But she was now head of her own section, with her own assistants and doctoral students, working independently of Hahn. She had many friends. Initially, her research concentrated on the study of beta radiation (fast-moving electrons) emitted by atomic nuclei; in 1921 she visited Copenhagen for the first time to describe her work at Niels Bohr Institute for Theoretical Physics. And in 1922, rather belatedly, she was appointed as the first 'privatdozentin' (female privatdozent) in Prussia, and only the second in Germany, entitling her to lecture at the university. This led to an incident that she dined out on for years. As part of the formal process involved, Meitner had to write a thesis, and she chose the subject 'The Significance of Radioactivity for Cosmic Processes'. A publisher, doubtless confused by her gender, asked for permission to reprint her essay on 'The Significance of Radioactivity for Cosmetic Purposes'. Hahn later recalled in his autobiography that one of his colleagues commented: 'it is incredible what one gets to be a privatdozent these days!' Clearly, Germany at large was not yet quite ready for female physicists.

In spite of her new status, Meitner was still, in her mid-forties, living like a student in a succession of rented rooms when the hyperinflation of 1923–24 wrecked the German economy. Her salary in August 1923 was several billion marks, but since a single egg cost two million that didn't go far. She was saved from the worst consequences by being invited to give a lecture tour of the Netherlands, where she was paid in hard currency and was able to send food parcels back to her friends.

The economy improved by the end of 1924, but Lise's mother died in December, weakening her links to Vienna. Berlin was now completely her home, and the prospects seemed good. She was established as one of the world's leaders in the experimental study of nuclear processes, with what seemed to be a secure position for life. It seemed to become even more secure in 1926, when, alongside her post at the KWI, she was made a professor at the University of Berlin – the first female physics professor at any German university.* As the icing on the cake, in 1927 her nephew Otto Frisch, the son of her sister Auguste, came to work in Berlin. He was a 23-year-old experimental physicist with a fresh doctorate from Vienna, and he got his new post without any help from his aunt, who declined to give him a reference. This was a year after Erwin Schrödinger and Werner Heisenberg had independently come up with their theories of quantum mechanics, and the same year that two groups of physicists independently proved that electrons could behave like waves. It was the golden time for experimental physics.

In the years up to 1933, the work of Meitner and her team covered just about all of the topics in the burgeoning field of experimental nuclear physics; but the one that led to her greatest contribution to science began with the discovery by James Chadwick, in Cambridge in February 1932, of the neutron, the neutral counterpart to the positively charged proton. This quickly led to the realisation that the nucleus of an atom consists of a ball of protons and neutrons held together by a strong force (called, logically enough, the strong nuclear force) surrounded by a cloud of electrons (one for each of the protons in the nucleus) following the rules of quantum mechanics. But that was just the beginning of the story.

The new understanding of atomic structure not only revealed that the chemical properties of an element depend on the number of protons (the atomic number), and therefore the number of electrons, but that atoms of the same element (the same number

* And she had, we are sure you will be relieved to know, at last moved into a proper apartment.

of protons) could exist in different varieties with different numbers of neutrons, and therefore slightly different mass. These variations on the theme are called isotopes. For example, there is a stable form of carbon called carbon-12, which has six protons and six neutrons in its nucleus, and an unstable form called carbon-14 – produced by reactions that we won't discuss here – which has two extra neutrons. A carbon-14 nucleus will decay by emitting an electron (beta ray), converting a neutron into a proton in the process and becoming a nucleus of nitrogen (specifically, nitrogen-14, with seven protons and seven neutrons).

Chadwick identified the neutron from experiments in which alpha particles (now known to be nuclei of helium) were used to bombard beryllium target atoms, which produced a powerful radiation that he explained as a stream of neutrons.* In modern terminology, the combination of helium-4 and beryllium-9 produces a nucleus of carbon-12 with a 'spare' neutron ejected at high speed. Meitner and her assistant Kurt Philipp quickly reproduced these results, and by measuring the energies involved in various experiments concluded that the mass of a neutron is a little more than 1.0056 times the mass of a proton; the modern estimate is 1.0087. With neutrons now identified, they could be used to bombard nuclei in the same way that alpha particles were used to bombard things like beryllium. When Meitner and Philipp learned that a team in Cambridge had made boron and alpha particles by bombarding nitrogen with neutrons (neutron plus nitrogen-14 makes

* For simplicity, we express this kind of interaction as if a single atom (or nucleus) is being targeted by a single particle; in reality, the stream of particles bombards a solid or liquid target containing many atoms and interacts with a few of the nuclei. The chemistry then comes in to extract the products of any interactions and identify them, while the physicists study the radiation produced. Then they have to explain what is going on!

Lise Meitner, with Otto Hahn in 1912

Public domain via Wikimedia Commons

boron-11 and helium-4), they not only reproduced the results but found a similar reaction in which a neutron plus oxygen-16 produced carbon-13 and helium-4.

In another series of beryllium–polonium experiments, the team also confirmed the existence of the positron, a positively charged counterpart to the electron, which had first been identified in cosmic rays in 1932. They studied the tracks of the various particles involved in the reactions using a cloud chamber, a sealed container in which the fast-moving particles left trails of bubbles. In a magnetic field, the charge on the particles makes them follow a curved track; some of the tracks attributed to electrons bent to the left, and some to the right, because electrons and positrons have opposite charge. Crucially, as well as being the first team to identify positrons from a laboratory source, they showed that positrons always appear paired with electrons. The result was reported in March 1933.

But this work was completed under a cloud. On 30 January 1933, Adolf Hitler was sworn in as chancellor of Germany. In order to disentangle the politics from the science, we will look briefly at Meitner's life under the Nazis before returning to the scientific story of the road to nuclear fission.

To some extent, although she was Jewish, Meitner was protected by being an Austrian, and not a citizen of the Third Reich. One of the first acts of the new regime was to introduce a law barring all undesirables, including anyone who had one or more Jewish grandparents, from holding a government appointment. This included university staff, but apart from her Austrian citizenship, Meitner was in a peculiar position because the KWI was funded by industry, and she was not technically a government employee. Her associate status as a university professor was cancelled in September, on the grounds that she was '100% non-Aryan', but this did not affect her salary or work at the KWI.

As well as the horrors inflicted on the population at large, large numbers of Jewish scientists were dismissed from their posts, including Otto Frisch, who had been working in Hamburg since 1930; the lucky (or wise) ones left Germany, taking what little they could – Frisch went first to London then on to Copenhagen. Einstein, who was in America at the time, never returned to his homeland. A few non-Jews made ineffectual protests; some, such as Schrödinger, left because they saw which way the wind was blowing; but many just kept their heads down and waited for the storm to blow over. Hahn, who liked to think of himself as apolitical, was also in America, but he returned in the summer of 1933 and cooperated with the Nazis in 'cleansing' his section of the KWI of Jews in order to keep it functioning. Meitner made no such concessions, but decided to stay and continue her work, which could not easily be started again somewhere else. However, like Hahn, she was given a Nazi Party member as her 'assistant' to make sure she didn't step out of line.

In spite of the 'help' of these party members, Hahn and Meitner were able to take on as an assistant a young chemist called Fritz Strassmann, who had seen all other options closed to him because he refused to join any Nazi-affiliated labour organisations, let alone the Nazi Party. In 1936, a group of scientists including Planck and Heisenberg tried unsuccessfully to have the Nobel Prize awarded to Hahn and Meitner, in the hope that this would give them (especially her) some protection by raising their status internationally. Throughout this period, though, right up until 1938, Meitner was even allowed to travel to scientific gatherings beyond the borders of the Third Reich, including trips to Brussels, Leningrad and Copenhagen. But then Germany absorbed Austria into the Reich, and the writing was on the wall – just as her work with Hahn and Strassmann was reaching a climax. It is time to get back to the science.

Meitner got up to speed with the new discoveries concerning the neutron and the positron at a meeting called the Solvay Congress, held in Brussels in October 1933. Just three months later, at the beginning of 1934, Irène and Frédéric Joliot-Curie, working in Paris, discovered that it was possible to make artificial radioactive isotopes. When they bombarded nuclei of aluminium with alpha particles, the process produced an isotope of phosphorus plus a neutron. This isotope, P-30, then emitted a positron and was converted into a stable isotope of silicon, Si-30.* Enrico Fermi's team in Rome followed this discovery by making artificial radioactive isotopes of a variety of elements, using a beam of neutrons instead of alpha particles. Meitner confirmed the results from Rome with her own experiments, and was particularly excited by Fermi's discovery that bombarding the heaviest known element, uranium (element 92 in terms of its number of protons), seemed to produce new radioactive elements – presumably some heavier than uranium, elements 93 and 94. She suggested to Hahn that they should renew their collaboration to investigate this phenomenon, with her doing the physics and him responsible for the chemical analysis of the substances produced by bombarding uranium with neutrons.

Before starting work on this project, they both attended a conference in Leningrad celebrating the centenary of the birth of Dmitri Mendeleev, credited with inventing the periodic table of the elements. When the uranium experiments began, however, Meitner and Hahn, like everyone working in the field at the time, were led up the garden path by assuming that the uranium interactions worked in the same way as interactions involving lighter elements. All those neutron bombardment experiments produced only small changes in the nuclei of the target atoms, adding or

* See Chapter Five.

subtracting one or two protons or neutrons. It was also natural to assume, from the understanding of the way electrons determine the chemical properties of an atom, that the 'transuranium' elements would be chemically similar to a lighter group of substances known as transition elements. Unfortunately, the chemistry seemed at first to confirm this. Strassmann, who had joined the team in 1935, was Meitner's main collaborator at this time, as Hahn was often distracted by administrative duties.

By the spring of 1937, it was clear that there was something wrong with the assumptions. In one example, there seemed to be a decay process that involved a single nucleus spitting out five electrons one after another, each one converting a neutron into a proton, so that the atomic number kept increasing even though the mass was unchanged. In a paper published that year, Meitner concluded that it was 'very difficult to reconcile [their results] with current concepts of the nucleus'.* Stuck up a dead end, the team took a break from the uranium work and carried out some experiments with the slightly lighter element thorium, atomic number 90. Irène Joliot-Curie, who was also investigating the supposed 'transuranium' elements, had suspected that the lighter element lanthanum-57 might be one of the products in the neutron-induced interactions, but when their own chemical analysis failed to find this isotope Meitner and Hahn dismissed the claim (incorrectly) as a mistake. That was where things stood when Germany absorbed Austria into the Reich in the spring of 1938, putting Meitner in immediate peril.

It soon became clear that Meitner had to leave Germany. One consequence of the Anschluss, however, was that Meitner's Austrian passport was no longer valid, and she had no legal papers

* *Zeitschrift für Physik*, volume 106, pages 249–70, 1937.

allowing her to travel outside what was now the Greater Reich. At the same time, many countries that had welcomed refugees throughout the previous years were beginning to feel overwhelmed, and tightened their restrictions on entry. Ruth Sime, in her excellent biography of Meitner, has described how over the weeks that followed, Meitner's options narrowed down to one – an invitation from Manne Siegbahn, a Nobel Prize-winning physicist who had recently been made the director of a new physics department of the Nobel Institute of the Royal Swedish Academy of Sciences, to join his institute in Stockholm, where the authorities were willing to allow her to work. On 12 July 1938, Meitner, using her Austrian passport, crossed the border into Holland without incident on a train, using a quiet line where the border guards at the sleepy crossing, unaware of her identity, let her pass without incident. She had with her a few clothes in a small suitcase, ten Marks in cash, and a diamond ring that Hahn had inherited from his mother and gave her to sell. From Holland, she flew to Copenhagen, nervously crossing German airspace, then travelled on to Sweden, where she arrived at the beginning of August. She officially joined Siegbahn's institute on a one-year appointment, on the understanding that it would be extended. But it soon seemed that the prospects of doing any worthwhile work were remote.

Siegbahn's institute was so new that it hardly had any equipment, and he was preoccupied with getting a particle accelerator installed and bringing everything up to speed. Meitner, who was coming up to her sixtieth birthday, was pretty much left to her own devices, in a strange country where she didn't speak the language, with little equipment and no assistants. The one (big) advantage of her situation was that she was safe from the Nazis. Under the circumstances, she could hardly be expected to come up with any important new ideas. A quiet retirement beckoned. But any such thoughts were far from her mind. As the news of more experimental results from Berlin and Paris came in, by the end of the year she had produced her greatest theoretical insight.

Both Curie's team in Paris and the Berlin team were grappling with the puzzle of a highly radioactive element with a distinctive radiation profile that was produced when uranium was bombarded with relatively slow-moving neutrons. Each nucleus of this form of uranium, U-238, is made up of 92 protons and 146 neutrons. The mystery product of the neutron bombardment had some properties similar to radium, but with other elements in the mix. Hahn suggested that when a neutron interacted with a U-238 nucleus it produced a nucleus of thorium (Th-235) plus an alpha particle, and that the thorium promptly spat out another alpha particle to become radium (Ra-231). This was a counsel of despair – no such double-alpha process had been seen before. But this was more or less where things stood in November 1938, when Hahn and Strassmann sent off a paper reporting their results to the journal *Naturwissenschaften*. Of course, Meitner's name should have been on the paper as well, but Hahn knew the risks of being formally associated with a Jew. But by each of them quietly arranging to visit Niels Bohr in Copenhagen in the middle of the month, Meitner and Hahn were able to get together and discuss the situation, without even Strassmann being aware that they had met. At this meeting, Meitner urged, or instructed, Hahn to carry out a more rigorous set of tests for the presence of radium. She insisted that the slow neutrons involved in the experiments simply did not have the energy to chip two alpha particles off the uranium. They needed a new explanation.

Other researchers seem to have been less worried about these anomalies. On 10 December 1938, Enrico Fermi was awarded his Nobel Prize in Stockholm, and spoke confidently about the discovery of elements 93 and 94, which his team had named ausonium and hesperium, and the probability of elements 95 and 96, as yet unnamed. He then took his family to America, rather than returning to Fascist Italy. Just nine days later, however, Hahn wrote to Meitner with the startling news that an intense chemical analysis of the products of the uranium experiments had revealed no trace of radium, but had found barium. But barium had an atomic weight of only 137 (Ba-137). Where could it have come from? On the day she received that letter, 21 December, Meitner replied that even the assumption of a 'thoroughgoing breakup' of the uranium nucleus could not be dismissed: 'one cannot unconditionally say: it is impossible'.* That is where things stood when Otto Frisch came to Sweden to spend Christmas with his aunt and some friends in the coastal town of Kungälv.

At this point, Meitner had one of those inspired insights that look obvious with hindsight but represented an entirely new way of thinking at the time. If a uranium nucleus containing 92 protons broke apart to produce a barium nucleus containing 56 protons, then what would be left behind? It had to be a nucleus containing 36 protons – which would be krypton. And how could the uranium nucleus split into two? Talking this over with Frisch, she picked up on an idea proposed by George Gamow and developed by Niels Bohr, which treated the nucleus as behaving in many ways like a drop of liquid. If the uranium nucleus was like a wobbling drop of liquid, then the incoming neutron could break it into two unequal droplets, perhaps with a few neutrons being splashed out as well.

The key discussion with Frisch took place on a snowy walk, when they sat down on a tree trunk and 'started to calculate on scraps of paper':† When the two drops separated they would be

* Letter in archive at Churchill College, Cambridge, quoted by Sime.

† This description comes from Otto Frisch's memoir What Little I Remember (Cambridge UP, 1969), and may be slightly coloured but is too delicious to leave out.

driven apart by their mutual electrical repulsion and would acquire very high energy, about 200 MeV in all; where could that energy come from? Fortunately, Lise Meitner remembered how to compute the masses of nuclei from the so-called packing fraction formula, and in that way she worked out that the two nuclei formed by the division of a uranium nucleus would be lighter than the original uranium nucleus by about a fifth the mass of a proton. Now, whenever mass disappears energy is created, according to Einstein's formula $E = mc^2$, and a fifth of a proton mass was just equivalent to 200 MeV. So here was the source for that energy; it all fitted!

Meitner sent news to Hahn and wrote up their idea for a joint paper with Frisch, in which at his suggestion they applied the word 'fission', consciously borrowed from biologists' use of the term to describe a cell dividing in two. He went back to Copenhagen and carried out experiments to measure the recoil of the nuclei involved, which confirmed Meitner's theory, and brought Bohr up to date with all the news. Frisch wrote this work up separately. Hahn and Strassmann, now on the right track, began a fresh chemical analysis of the products of the neutron bombardment of uranium, including a search for krypton. The supposed 'transuranium' elements turned out to be other fragments produced by fission, although genuine transuranians have since been manufactured. Once again, for political reasons Meitner's name did not appear on their papers. But as Strassmann later wrote:*

> Her [Meitner's] initiative was the beginning of the joint work with Hahn … and she was bound to us intellectually from Sweden [through] the correspondence … [She] was the

* Quoted by Sime, from the German original.

> intellectual leader of our team, and therefore she belonged to us – even if she was not present for the 'discovery of fission'.

The joint paper by Meitner and Frisch appeared in *Nature* on 11 February 1939, and his solo paper a week later. Serendipitously, Niels Bohr had left on a visit to the United States on 7 January and was able to spread the news of the discovery of nuclear fission, with full credit to Meitner, even before those papers had crossed the Atlantic. As research into nuclear fission took off, Meitner was in Sweden, frustrated by her inability to join in because of her lack of equipment. In July 1939, she visited Cambridge, where she received a warm welcome at the Cavendish Laboratory and the offer of a three-year appointment linked to a fellowship at Girton College. She decided to return to Sweden to sort things out there and take up the position in the summer of 1940, but the outbreak of the Second World War in September 1939 effectively trapped her in Sweden, and brought a decline in her productive career – although it is unlikely that she would have been very productive, even at the Cavendish, in her sixties. She should have enjoyed a semi-retired life as one of the acknowledged pioneers of nuclear physics, and so she did, except for one extraordinary omission.

In 1945, the Royal Swedish Academy of Sciences was catching up with unfinished business after the war, and announced on 16 November that, among other awards, the 1944 prize for chemistry, held over because of an argument among the Committee, had gone to Otto Hahn and the 1945 prize for physics had gone to Wolfgang Pauli. The citation for Hahn's award specifically said that it was 'for the discovery of the fission of heavy nuclei', although the very term fission had been introduced in this context by Meitner and Frisch. The reason for the delay in announcing Hahn's prize was that, even by the often bizarre decisions of the Nobel Committee, this was astonishing. So much so that when the award was originally proposed it had provoked a furore in the scientific community.

It would have made sense to give the chemistry prize jointly to Meitner and Hahn, but if only one member of the team were to be honoured it ought to have been Meitner, who as Strassmann acknowledged was the team leader. It might also have made sense to give the chemistry prize to Hahn and Strassmann and a share of the physics prize to Meitner. But to ignore her entirely was outrageous.

Faced with a storm of protests, in 1944 the Academy made the unprecedented decision to reconsider, while holding the award over for a year; but the all-male committee stuck by their original decision. Letters from the physicist Oskar Klein to Niels Bohr (now in the Bohr archive in Copenhagen) reveal that after a stormy meeting the original decision was confirmed by a narrow majority, so the award did indeed go to Hahn alone. After the deliberations of the committee were made public, fifty years later, Max Perutz, himself a Nobel laureate, commented:*

> the documents leading to this unjust award now reveal that the protracted deliberations by the Nobel jury were hampered by lack of appreciation both of the joint work that had preceded the discovery and of Meitner's written and verbal contributions after her flight from Berlin.

And in October 2020, the editors of the journal *Physics World* said: 'If ever there was a discovery that should have won its instigators a Nobel Prize in Physics, nuclear fission is surely it.' They concluded: 'The 1944 Nobel Prize in Chemistry represents a strong challenge

* A Passion for Science, *New York Review of Books*, 20 February 1997.

to anyone who claims that the Nobels are fair or reflective of how collaborative science works.'

Because of the prestige associated with the Nobel Prize, in popular accounts Meitner began to be referred to as Hahn's co-worker, with the implication that she was his assistant. And adding insult to injury, when Hahn came to Stockholm to receive the prize in 1946, in his official speech of thanks he made only a brief mention of her contribution. From then on, he behaved as if she really had just been his assistant.

Others knew better. Meitner was nominated 49 times for either the physics or chemistry prize, but it was one of the few scientific honours she did not receive. She stayed in Sweden, continuing to contribute to nuclear physics theory from 1947 at the new atomic research section of the Royal Institute of Technology (where she did have assistants), and only officially retiring in 1954, when she was 75. In 1960 she moved to England, where Otto Frisch was now a professor at the University of Cambridge, to be near his family. In 1966, the US Department of Energy jointly awarded her, Hahn and Strassmann its prestigious Enrico Fermi Award, the first time it was awarded to a woman, citing her 'pioneering research in the naturally occurring radioactivities and extensive experimental studies leading to the discovery of fission', which is more or less what the Nobel Committee should have done two decades earlier.

Meitner lived just long enough to see this belated recognition; she died on 27 October 1968, in her ninetieth year. Hahn had died a couple of months earlier. In 1994, element number 109 was formally named meitnerium in her honour. As yet, there is no element officially called hahnium.

CHAPTER 3

Emmy Noether (23 March 1882–14 April 1935)

Found the mathematics that underpins particle physics

Emmy Noether was a near contemporary of Lise Meitner, came from a similar Jewish-German background, and had to overcome similar obstacles in order to follow a career in science. But in spite of the parallels, her story is worth telling here because she was not only a female mathematician – a rarity in itself – but was described by Albert Einstein as the most important woman in the history of mathematics. How could we leave her out?

Amelie Emmy Noether was born on 23 March 1882 at Erlangan, in Bavaria. Her given names were indeed Amelie Emmy – Emmy was not, as some accounts suggest, a contraction of Amelie, but the forename she actually used (the 'Rufname', as it was called). She was the oldest child of Max and Ida Noether, and would have three younger brothers. They were part of a small Jewish community, only about 200 people, in the city, which had been closed to Jews until 1861. The family was well off; Ida came from a wealthy family, and Max was one of two professors of mathematics at Erlangen University. His colleague Paul Gordan was also Jewish, a unique situation for mathematical faculties in Germany at the time. Max was a leading theoretical mathematician whose fame spread far beyond Erlangan, and he clearly inspired his daughter's love of the subject, although she had obstacles to overcome before those dreams could be fulfilled.

Like Meitner, Emmy was forced to leave school at an early age, and obtained a qualification to teach French and English at female colleges. But her parents paid for her to have private tuition in mathematics and other subjects, while she was allowed to attend lectures at the university as a visitor. She also studied with her brother Fritz when he was preparing for the university entrance examination. Emmy then obtained permission to take the examination in July 1903, when she was 21, although she had never attended the Gymnasium and women were still barred from entering most German universities. She duly passed – just two months before the Bavarian ruler, Prince Luitpold, gave women the right to attend university if they had this qualification.

In spite of this, Emmy chose to attend classes at the University of Göttingen, in Prussia, where she was only allowed to audit the lectures with the permission of the lecturers. Starting in the autumn of 1903, in her haste to make up for lost time she attended lectures by the eminent mathematicians David Hilbert, Felix Klein and Hermann Minkowski, and the mathematician-astronomer Karl Schwarzschild, each of whom would later make significant contributions to relativity theory. Possibly because of overwork, she became ill at the end of the year and returned to Erlangan, where she recovered and enrolled for a degree course at the university. Starting in the autumn of 1904, she was one of only two women admitted that year. Following the usual path through a German university at the time, she emerged in December 1907 with a doctorate based on some conventional mathematical work supervised by her father's colleague Paul Gordan. But that was as far as she was officially allowed to go.

In order to be paid to teach, the traditional path would be to go through the formal process known as habilitation, involving submission of another thesis, to become a privatdozent. But women were prohibited from habilitation. All she could do was assist her father and Gordan in their work, teaching courses in their names and even supervising doctoral students, with no remuneration. But she also published a series of mathematical papers, which began to get her name noticed. It was during this period of her life that Noether began to work on what became her major contribution to mathematics, and although it took time for these ideas to be developed and their importance to be appreciated, this seems a good place to explain briefly what it was all about, while we leave Noether marking time in Erlangan.

To many people, the word 'algebra' conjures up images of equations involving things like x and y, and memories of trying to solve them to find the values of x and y. This kind of arithmetical manipulation goes back more than 4,000 years, to Babylonian times. If the equations only involve xs and ys (and other 'plain' numbers), they are said to be linear. Equations with squares in (x^2 and so on) are quadratic, and there are equations involving cubes and higher powers – higher order equations. Each kind of equation can be solved using its own specific set of rules. But in the nineteenth century, mathematicians found that there is a deeper way of describing algebra, an underpinning set of principles that you can use to solve equations of any order. Later in the same century, mathematicians were investigating different kinds of geometry. The basic geometry we learn in school involves triangles and other things drawn on a flat surface, where, for example, the angles of a triangle always add up to 180 degrees. But there are other kinds of geometry, such as the properties of shapes drawn on the surface of a sphere, which follow different rules – on a sphere, among other things, the angles of a triangle always add up to more than 180 degrees. It turned out that these different geometries could also be described by an underlying set of principles that apply to all geometries. Even better, this is the same set of principles that underlies the techniques of solving algebraic equations. These underlying principles – which also apply in some other circumstances – became known as group theory. By the early twentieth century, other mathematical puzzles had also been found to share certain basic principles (not all the same as group theory), and the study of these underlying principles became known as modern algebra (later, abstract algebra), which it certainly was, compared with the Babylonians.

There is an analogy sometimes used to give an idea of how valuable this approach can be. Dogs come in many different varieties, and at first sight it is hard to believe that a chihuahua and an Irish wolfhound are members of the same species. But all dogs share basic characteristics and are built using the same principles. A vet doesn't have to learn in detail about each variety, because they can learn the underlying principles and apply them to whatever dog they happen to be treating – even if it is a breed they have never seen before.

Noether was fascinated by modern algebra, not so much for its power at solving practical problems, but because of the deep insights it provided into the nature of mathematics and the world. It was almost incidental that her work turned out to have hugely important implications for physics. In the 1920s, she became the leading proponent and teacher of modern algebra, and the definitive textbook on the subject, written by one of her protégés, was based on her lectures. But that was only possible after a long struggle, when she did obtain permission to teach, in Göttingen after the First World War. And it was partly thanks to the war that she became established in Göttingen. It is time to pick up the story from her days as an unpaid and unofficial assistant in Erlangan.

What would turn out to be a key event in Noether's life occurred early in 1913. Her father's friend, and her former supervisor, Paul Gordan, had just died, and Max Noether was asked to write his obituary. As part of his preparation for this, Max decided to visit Felix Klein, who had been one of Gordan's collaborators, in Göttingen. Emmy, now in her early thirties and with a modest but growing reputation of her own, went with him to take the opportunity of discussing mathematics with Klein and his colleagues. She made a deep impression on them. Klein had written a book on fifth-degree polynomials ('quintic' equations, which include x^5 and lesser powers of x), and after discussing such equations with Emmy he felt that she knew the subject better than he did. He later wrote that 'she not only had full command of one of my earlier research fields, the theory of quintic equations, but that she could inform me about several details that were new and gave me great satisfaction'.*

The reason Klein wrote that letter, in 1915, was to support Noether's efforts to habilitate in Göttingen. She had moved there after the outbreak of the First World War because Hilbert and Klein, whose young colleagues were being called up to fight, needed a reliable assistant who knew mathematics and was not going to be called up. Who better than the young woman who had impressed them a few months earlier? But even with such support, Noether's road to habilitation was long and far from straightforward.

Noether arrived in Göttingen in the spring of 1915† and almost immediately (on 20 July) applied for habilitation, although women were still forbidden by law to habilitate. The formal procedure involved the university faculty first agreeing to put the applicant's name forward, and then the appropriate ministry approving the

* Letter quoted by Rowe and Koreuber.

† Her mother died suddenly a few weeks later, but there is no record of any immediate impact of this on Emmy's life.

recommendation – the second step was usually just a formality. At that time, the mathematics and natural sciences department at Göttingen was part of the faculty of philosophy, and by and large the mathematicians and scientists supported the application, while the humanists opposed it. Everyone agreed that Noether was an above-average candidate and that there would be no problem recommending her if she were a man. Even some of her supporters, however, were at pains to emphasise that she was a special case, not typical of her sex, and that if she were allowed to habilitate this ought not to open the doors to other women.

The philosophy faculty voted by ten to seven in support of the application, with two abstentions. The debate then went to the faculty of the entire university, where, after heated arguments during which Hilbert commented: 'I do not see that the sex of the candidate is an argument against her admission as privatdozent. After all, we are a university, not a bathhouse', a vote was taken on a proposal to urge the minister of education to reject the nomination being put forward by the mathematicians. This vote was tied at fourteen all, and only defeated by the casting vote of the dean. The mathematicians immediately drafted their formal request to the minister, which was submitted on 9 December. Significantly for our story, this was a few weeks after Einstein completed his general theory of relativity, about which he had been corresponding with Hilbert.

Even at this late stage, though, there were still objections from within the philosophy faculty. The losers in the debate formally put in a minority report, which gives us a clear idea of the problems women faced at the time. Among other things, it said that a woman 'is unsuitable for regular instruction of our students because of the phenomena connected with the female organism' (whatever that may mean), and that it would be unfair to let women habilitate because that might lead them to imagine they could go on to become professors, which was obviously impossible, so their expectations should not be raised. Johannes Hartmann, an astronomer, added that it would be dangerous if 'mentally superior women were more and more removed from family life' because society would lose their superior offspring if they did not marry and have children. And, he noted, while it had been a great help to have women replace, on a temporary basis, men who had gone off to war, if they were allowed to habilitate, 'our private lecturers returning from the field … would surely greet such competition, which had emerged during their absence, with very mixed feelings'. This point rather overlooked the fact that before the war Göttingen had struggled to find any male candidates suitable as lecturers in mathematics.*

The request to the Ministry of Education was rejected, but the minister offered Hilbert a compromise. Noether would be allowed to teach courses in Göttingen under Hilbert's name, as his 'assistant'. Of course, Hilbert had nothing to do with these lectures, which began in the autumn of 1916. The following summer, when the university heard that the University of Frankfurt had put out feelers about the possibility of Noether going there to habilitate, the mathematicians in Göttingen asked the ministry to reconsider her case, but were assured that Noether would not be allowed to habilitate there either.

It was against this background that, in 1915 and 1916, Noether laid the groundwork for what is now regarded as her greatest work, which we shall explain shortly. But first, we will complete the story of her habilitation.

Noether's move to Göttingen had been prompted by the outbreak of the First World War, and her elevation to the status of privatdozent was one of the side effects of the chaos in Germany

* English translation from Rowe and Koreuber.

at the end of the war. There was another side effect of that chaos. In 1919, Noether joined the Independent Social Democratic Party (USDP, from its initials in German), remaining a member until 1924. Although she was not politically active, this was sufficient for her to be labelled as a Marxist in later years. Back in 1919, however, her leftist sympathies were shared by most of the intelligentsia, and posed no obstacle to her renewed fight for habilitation. By the beginning of 1919, the first year that women in Germany were able to vote, the Ministry of Education was itself headed by two socialists, one of whom, Adolph Hoffmann, was a member of the USDP, and when the entire philosophy faculty at Göttingen voted unanimously to revive Noether's application to habilitate and it went back to the Ministry, it was formally approved and she gave her inaugural lecture on 4 June that year. At last she could become a privatdozent, get paid for lecturing and, as Hartmann had feared, start up the academic ladder. She had still been treated as a special case, but in 1920 the law preventing women from habilitation was finally abolished.

Noether's inaugural lecture and habilitation thesis were based on her breakthrough work, which had resolved a problem that David Hilbert and Felix Klein thought they had identified with Einstein's general theory of relativity in 1915, soon after Einstein had unveiled his masterwork. At first sight, it looked as if the general theory violated one of the fundamental laws of physics, the conservation of energy. This says that energy can never be created or destroyed, but only converted from one form to another. For example, a moving car has energy due to its motion – kinetic energy – and when you use the brakes to bring the car to a halt this is converted into heat – thermal energy – in the brakes, and dissipated out into the world at large. The general theory is a theory of gravity, and the situation is complicated because of the relationship between matter, space and time (or spacetime), which lies at the

Emmy Noether, c. 1900
Public domain, via Wikimedia Commons

heart of the theory. Although the theory as formulated by Einstein did, in effect, contain an equation which could be interpreted to refer to the conservation of energy, Hilbert and Klein noticed that this was equivalent to an equation which says $x - x = 0$, which is true but doesn't tell us anything about x. This left open the possibility that energy is not really conserved according to the general theory, and they roped in Noether, not long after she had arrived in Göttingen, to see if this was really the case.

In a brilliant piece of mathematical analysis, Noether was able to show that energy must be conserved, using the general theory or any other theory of gravity, provided that the laws of physics stay the same for all times – that they are invariant. This rule works both ways – if energy is conserved, then the laws of physics must be invariant with respect to time. This drew praise from the great man himself. On hearing the news, Einstein wrote to Hilbert saying: 'Yesterday I received from Miss Noether a very interesting paper on invariant forms. I am impressed that one can comprehend these matters from so general a viewpoint.'

But this was just the beginning. Noether was immediately able to extend her insight to cover many other laws of physics, linking them to so-called symmetries in the equations. In this sense, time invariance is a symmetry because it doesn't matter where (or when) you are in time, the rules are the same. There are two kinds of symmetry, which we can understand by looking at simple geometry rather than at the equations. If you take a square, and rotate it by 90 degrees, it looks just the same as before. This is a kind of symmetry. But if you rotate the square by 45 degrees it looks different – like a diamond. So this kind of symmetry is called a discrete symmetry, because you have to go in discontinuous steps to restore the original appearance. By contrast, if you have a circle you can rotate it through any angle you like, and it still looks the same. This is called a continuous symmetry. What has become known as Noether's theorem proves that in nature the existence of any continuous symmetry is always linked with a conservation law. You can use the symmetry to work out the conservation law, or you can use the conservation law to reveal the underlying symmetry. Some simple examples show what a powerful insight this was.

In all these examples, we ignore friction, so we can imagine, for example, a hockey puck sliding endlessly at a constant speed over a smooth sheet of ice.* We all know from experience about the law of conservation of momentum. If the puck strikes an identical stationary puck head on, it will stop, while the second puck moves off in the same direction at the same speed, carrying the momentum of the original puck. If the first puck strikes the second puck a glancing blow, they will both move off in different directions at different speeds (less than the speed of the original puck), sharing the momentum of the original puck. This property – conservation of momentum – turns out to be linked to a symmetry of space, which says that the puck moves the same way wherever it is on the icy surface.

Sticking with icy analogies, we have all seen how the spin of an ice skater changes as they pull their arms in or stretch them out. The spin is faster when the arms are tugged in tight. This is a result of the conservation of angular momentum, because a more compact object spins more rapidly than an extended object with the same angular momentum. And it is linked to a rotational symmetry. If you do an experiment on a lab bench, then rotate the bench by an arbitrary amount and do the experiment again, you will get the same result. The conservation of energy, which got Noether involved in all this in the first place, is related to a time symmetry,

* Of course, physicists can do the 'proper' calculations including friction, but there is no need for us to go into that here.

which says that if you do the experiment on a Thursday you will get the same answer as you would if you did it next Sunday.

All this helped to establish Noether's theorem, but didn't tell us anything new about the world. The power of the theorem is that it can be applied in less everyday situations to explain other situations and predict the behaviour of particles at the quantum level.* The symmetries involved cannot be visualised in a simple way like space symmetry, but they are described in mathematical terms and linked to physical phenomena. Physicists can identify particles which seem to behave in accordance with a particular symmetry, then use that symmetry to predict the existence of other particles, such as quarks and the Higgs boson. Noether's theorem is one of the most powerful tools physicists have for understanding the behaviour of quantum particles and fields. Two leading physicists of the twenty-first century described it as 'one of the most important mathematical theorems ever proved in guiding the development of modern physics, possibly on a par with the Pythagorean theorem'.†

But that isn't Noether's only contribution to our understanding of the subatomic world. Although she had developed her theorem in 1916, it wasn't published until 1918. Just three years later, in 1921, she published an important paper on what mathematicians call ring theory, although this does not refer to the kind of rings you wear on your hand. 'Rings', to a mathematician, are sets of things which can be added together or multiplied by one another to make other members of the ring. The classic example is the set of integers 1, 2, 3, 4 and so on. It turns out that more complicated rings

* It also applies on a cosmic scale, to things like galaxies. It is literally a universal truth.

† Leon Lederman and Christopher Hill, in *Symmetry and the Beautiful Universe*, Prometheus Books, Amherst, 2004.

are associated with the behaviour of quantum entities, although we cannot offer you any simple examples like the application of Noether's theorem to the rules of momentum. However, more than a hundred years after she published those papers, her work is still at the cutting edge of theoretical physics.

After 1921, Noether continued to make important contributions to mathematics, but in rather esoteric realms which did not transform theoretical physics. (For a no-punches-pulled account of this phase of her work, see *Proving It Her Way: Emmy Noether, a Life in Mathematics* by David Rowe and Mechthild Koreuber.) Her life was not dramatically improved by her new status as a privatdozent: she was able to make a modest income from lecturing and was given the honorary title of associate professor in 1922, but this did not carry any salary. In the immediate post-war years, Noether spent a lot of time away from Göttingen looking after her father, whose wife had died in 1915, until he eventually died on 13 December 1921 at the age of 77. The economic collapse and runaway inflation hit her hard, but she just about scraped by, and lived in the same modest apartment from 1922 to 1933. Her life in this period revolved around mathematics, her own research (including visits to foreign universities) and teaching several PhD students, who enthusiastically took up her ideas and spread them as their own careers developed. Her lectures bore little relation to what most of us recognise by the term. She did not have prepared notes, but used the time for discussions with her students, working on important problems in mathematics. Some of her own significant ideas were developed in this way, and the lecture notes taken by her students provided the raw material for important textbooks written by other people. Her colleagues also attended her lectures, which were more like research seminars, and she was always happy when the results that came out of these discussions were published by others, including her students.

One of those students (she happened to be the first) is worth picking out for mention. Grete Hermann completed her doctorate in 1926, with a thesis on an aspect of ring theory which is still important today in applications involving computer algebra. Six years later, the mathematician John von Neumann published a book on the mathematical bases of quantum mechanics, in which he claimed to have proved that a straightforward way of understanding the quantum world, called hidden variables theory, was flawed. Von Neumann was so highly regarded that everyone accepted his word as gospel and abandoned this approach. But in 1935 Hermann found a flaw in Von Neumann's argument, a very simple mathematical slip, and published a paper pointing this out. Hidden variables theory was not impossible after all. The paper was totally ignored, and hidden variables theory remained taboo until John Bell independently found and pointed out the flaw three decades later. If Hermann's paper had not been ignored, the history of the development of quantum theory would have been quite different. It is impossible to be sure why her paper was ignored, but one factor may well have been that it was hard, at that time, to take seriously a young woman who was challenging the work of an acknowledged male genius. But that was the calibre of Noether's students.

For Noether, 1932 began well. She shared the highest German award for mathematics, the Ackermann–Teubner Prize, with the Austrian Emil Artin, of the University of Hamburg. Artin was a disciple of Noether's approach, and the year before they shared the award, the influential book *Modern Algebra*, written by the Dutchman B.L. van der Waerden (who had been a visitor at Göttingen in the mid-1920s) but based on lectures by Noether and by Artin, had been published. It must have seemed, as she turned 50, that Noether's position was secure. But by the end of the year the political winds of change were blowing hard, and nothing was certain.

In the familiar pattern, after the Law for the Restoration of the Civil Service was passed in April 1933, the Ministry of Education placed six of the Göttingen faculty, including Noether, on immediate 'leave of absence'. She was not only Jewish but also regarded as a Marxist, doubly damned in the eyes of the Nazis. In spite of an appeal to the Ministry backed by glowing testimonials to her mathematical genius and political soundness, there was, of course, no reversal of the decision, and her licence to teach was withdrawn in September. But Noether was luckier than many of her contemporaries. She received an invitation to spend a term at Oxford, and also for a one-year appointment (later extended to two years) at Bryn Mawr College in America. Expecting, like many people, that the political situation in Germany was only temporary, she accepted the American offer and crossed the Atlantic, arriving at the beginning of November 1933.

Bryn Mawr College, not far from Philadelphia and only a little further from Princeton, had been founded in 1885 for the education of women, and was particularly strong in mathematics – it was the only college in the United States where women could study for a PhD in the subject. The suggestion that Noether might be offered a post came from Solomon Lefschetz, at Princeton, who had met Noether in Göttingen in 1931. This idea was enthusiastically taken up by the college and led to an appeal for funding to the Rockefeller Foundation, which agreed to cover her salary as visiting professor of $4,000 – about $75,000 in today's money – for one year. This was the only time in her life that Noether would be appropriately rewarded financially for her work.

Noether was a popular and successful teacher at Bryn Mawr, and from February 1934 also travelled to Princeton once a week to lecture at the Institute for Advanced Study. With her funding at Bryn Mawr extended for another year, she took a holiday in the summer of that year to visit Germany, where she had kept her apartment and which she still regarded as home. But it soon became clear that the situation was much worse than she had realised, and even in Göttingen she was only allowed access to the library as a 'foreign scholar'. Under no remaining illusions, she made arrangements to give up the apartment and ship her belongings to the United States, even though at that time she had no guarantee of employment beyond 1935. Then, she left Europe for the last time.

Back at Bryn Mawr, the Rockefeller was persuaded to commit themselves to funding her for another two years. But the money was never needed. In April 1935, Noether went into hospital for a straightforward operation to remove an ovarian cyst. Everything went well, but after the operation she suffered a brain haemorrhage and died, on 14 April 1935. It seems likely that the weak spot in her brain might have given way at any time, although the stress of the operation could have been the trigger. In any case, her surgeon was convinced that without the operation the tumour would have killed her. She was just 53. In a letter to *The New York Times*,* Albert Einstein wrote:

> In the judgment of the most competent living mathematicians, Fräulein Noether was the most significant creative mathematical genius thus far produced since the higher education of women began. In the realm of algebra, in which the most gifted mathematicians have been busy for centuries, she discovered methods which have proved of enormous importance in the development of the present-day younger generation of mathematicians.

Today, we might well delete 'since the higher education of women began', but even Einstein was imbued with the sexism of his day.

* 5 May 1935.

CHAPTER 4

Inge Lehmann (1888–1993)

Discovered the inner core of the Earth

A year after Emmy Noether died, the Danish scientist Inge Lehmann reported the discovery that the Earth had a solid inner core. Lehmann had been born on 13 May 1888, at Østerbro near Copenhagen. But she died as recently as 21 February 1993, in her 105th year, and largely because of this her archive of personal papers only became available to researchers in 2015. The first historian to gain access to those papers, Lif Lund Jacobsen, found that they revealed new insights into Lehmann's life and the difficulties faced by female scientists in the first half of the twentieth century – which, as we mentioned in the introduction to this book, make her the archetypal example of a female scientist overcoming the obstacles placed before her by society (including, in this case, by her father). The following account is largely thanks to Jacobsen's analysis of that archive.

Inge Lehmann grew up in a family with academic connections and links to feminist activism. Her father, Alfred, had a degree in applied science, but became a psychologist, achieving the rank of professor at the University of Copenhagen in 1910. Her mother, Ida, was involved in the women's rights movement, and one of Ida's cousins became the minister of trade. Ida's younger sister, Signe, became an independent single mother and a school superintendent.

Inge and her own younger sister, Harriet, went to the first co-educational school in Denmark, where there was equal tuition for boys and girls. This was an isolated example; until 1903, in Denmark boys could take the necessary examination and move on to high school at fifteen, but girls had to wait until they were seventeen. Partly because of the difficulty women had in obtaining scientific posts, the teaching at the school was first class. Its founder, Hanna Adler, was one of the first women in Denmark to earn a Master's degree in physics, and she employed many women who had degrees but were excluded from teaching at boys-only high schools, let alone university. Adler was also the aunt of the future Nobel Prize-winning physicist Niels Bohr (born in 1885), whom Lehmann got to know.

Lehmann excelled at mathematics and physics, and graduated from high school in 1906. Her father initially encouraged her ambition to get a scientific degree, and after a brief spell as a private tutor she enrolled to study mathematics at the University of Copenhagen in 1907, when she was nineteen. Little is known about her life over the next few years. She lived quietly with her parents, studied hard and passed the first part of her degree examinations with flying colours in the summer of 1910. But rather than continuing to complete the second part of her degree, she moved on to Newnham College, in Cambridge – with Girton, one of only two colleges for women in Cambridge at that time – to study mathematics. There was strong competition for places at both colleges, and Newnham had a high reputation for maths, so this was a significant achievement. 'Study' is the key word here, however, because although women were allowed to attend lectures in Cambridge, and even to take examinations, until 1948 they were not allowed to actually receive degrees. Nor was it allowed for men to tutor the young women. All this was a considerable culture shock for Lehmann, which came on top of the intense study required for the course (the 'Maths Tripos'), which included astronomy and physics as well as both pure and applied mathematics. In her autobiographical memoir* she highlighted 'the severe restrictions inflicted on the conduct of young girls, restrictions completely foreign to a girl who had moved freely among boys and young men at home'.

The absurdity of the situation, and its impact on Lehmann, is highlighted by her interactions with Niels Bohr, who had just completed his PhD and came to Cambridge in the autumn of 1911 to work for a time at the Cavendish Laboratory. Before he arrived, he wrote asking Lehmann for advice on which lectures might interest him. She was happy to help, and also to assist him in finding a place to live. But once he arrived, things became more complicated. There were strict rules about students socialising with members of the opposite sex, which made it very difficult for female students to enjoy free discussions with their male counterparts about their coursework and so on. Specifically, Inge had to have a chaperone (female, of course) if she wanted to visit a man, or have him visit her. The identity of the chaperone had to be approved by the college authorities in advance. This applied to every meeting she had with Bohr, and at least once a planned get-together had to be cancelled because no suitable candidate was available. Things reached a head in December 1911. As Inge was going back to Copenhagen for Christmas, Bohr invited her to a festive dinner at his lodgings. By now wise (he thought) to the ways of Cambridge, he also invited a friend and his friend's sister, who could act as chaperone. Alas, the sister was also a student at

* See Bibliography.

Newnham, and was therefore not regarded as suitable. After an exchange of letters, Bohr invited some more guests, one of whom qualified as a chaperone in the eyes of the college.

After the dinner, Lehmann left Cambridge, intending to return in the New Year. But she never did. The reason for her change of mind had long been a puzzle – leading to speculation that she suffered a nervous breakdown – but the truth was uncovered by Jacobsen on her trawl through the archive.

The key reason why Lehmann did not go back to Cambridge was that her father refused to continue her financial support, and insisted that she got a job in Denmark, away from the stresses of academia. It is clear from the correspondence that is now available that she was very tired as a result of hard work, a classic example of a woman trying to prove that she was better than the men, although she was far from having a breakdown and would have been well able to carry on at Cambridge. But her father seems to have been influenced by the work of a Danish doctor, Leopold Meyer, who had published a series of papers claiming that too much intellectual study would make women ill, because their bodies were not suited to such work. In his letters to Inge, Alfred Lehmann mentions the 'sad examples of how it went with intellectually gifted women who wanted something more' and became 'half insane'. Of course, he does not give any case studies to back this up. Inge's response to this is not recorded, but it was clearly forceful, since in his next letter Alfred proposes that she stay at home until September 1912, then go back to the University of Copenhagen to complete her MA if she felt strong enough. He even promised that if that went well he would support her eventual return to Cambridge. As Alfred surely anticipated, though, none of this came to pass.

Instead of returning to the university, in the autumn of 1912 Inge started work in the actuarial department of an insurance company, where although the work was routine it at least involved some skill with statistics. Denmark remained neutral during the First World War, and she stayed in the job, living with her parents, for five years, although she was repeatedly passed over for promotion while her male colleagues progressed. In 1915, women were given the right to vote in Denmark, but that doesn't seem to have improved their employment prospects. As she approached her twenty-ninth birthday, Lehmann seems to have decided that enough was enough, and took drastic action to change her situation. First, in February 1917, she became engaged and quit her job, since married women were not expected to work. Almost immediately, she realised that she had made a mistake, broke off the engagement, and after a year without employment (unfortunately there are no records of her father's reaction), in August 1918 she returned to the University of Copenhagen where she collected her Master's degree in mathematics and physical science with first-class grades in 1920. A year later, her father died, and about the same time a law was passed in Denmark giving women equal employment rights, including access to university posts. These rights allowed women to take up academic jobs but did not guarantee them equal pay. Nevertheless, Lehmann was now able to work within the university system, free from any parental influence, although that also meant she had to pay her own way. After a year studying maths in Hamburg, funded by a modest scholarship, she joined the staff of the Actuarial Mathematics group at the University of Copenhagen in March 1923. Her salary was a modest 700 kroner – less than half that of her male colleagues.

In order to supplement her income, Lehmann took on a succession of temporary part-time jobs, including translation and editing work for Niels Nørlund, one of the mathematics professors. It is hard to see how she managed to fit this in, as her official duties included tutoring students, assisting with practicals, and the time-consuming job of marking. After intense lobbying by the head of the laboratory, Johan Steffensen, her salary was brought in line with that of the men in February 1925, but she soon managed to find an even better position.

As well as his role as professor of mathematics, Nørlund was the director of the Danish Geodetic Service, and in June 1925 Lehmann asked him if there was any chance of a research position. There was nothing available at the university, but the Geodetic Service was about to expand its activities by setting up new seismological stations in Copenhagen, southwest Greenland and west Greenland. He offered her the chance to supervise this work, in an administrative capacity, alongside her work at the Actuarial lab. In her memoir, she described how: 'I and three young men who had never seen a seismograph before were active installing [them] in Copenhagen and also helping to prepare the Greenland installations. I studied seismology at the same time unaided.'

She did all this so well that when everything was up and running, in January 1927, Nørlund appointed her to the full-time job of running the new network, and arranged for her to spend several months abroad (in Strasbourg, Prague and Darmstadt) studying seismology. The most important part of this education came from a month spent learning from one of the world's leading geophysicists, Beno Gutenberg, in Darmstadt. In the spring of 1928, the network was absorbed into the newly created Danish Geodetic Institute, with Nørlund as director, and in May (the month of her fortieth birthday) Lehmann, using the knowledge gleaned from her European trip, passed the examination to be awarded an MSc in geodesy. A few weeks later, she became the director of the seismology department at the Geodetic Institute, a post she held until she retired in 1953. In all that time, she was the only academic in the seismology department, with complete responsibility.

Inge Lehmann, 1932

Even Neuhaus, Public domain, via Wikimedia Commons

Lehmann was well aware of the faith Nørlund had placed in her. In November 1928 she wrote to thank him:*

> I could not have wished for anything better. I have earlier been concerned that I was asking too much when refusing to be satisfied with working in order to earn money, but sought a job in which I could really take an interest. In my work here I have … found more than I could ever have hoped. In return, I shall do my utmost. It is no small thing to have the opportunity and permission to use all one's strengths.

That faith would be amply rewarded.

Lehmann's duties included keeping the seismic monitoring equipment up to scratch, arranging building maintenance, selecting, training and overseeing the operators for the remote stations, and, crucially, interpreting every seismogram from all three sites and publishing the results of their earthquake measurements in a regular bulletin. She was allowed to use the data for her own research, provided she did it in her own time – although she was encouraged to concentrate on her proper job. Over the next few years, her familiarity with seismograms gave her an intuitive feel for their interpretation, which a colleague later described as 'a black art for which no amount of computerization is likely to be a complete substitute'.† It was this, and the location of her seismic network in Europe, that enabled her to identify the inner core of the Earth.

The location is important because Europe lies on the opposite side of the world to the seismically active regions of the globe in the South Pacific, so earthquake waves travel right through the

---

* Quoted by Jacobsen.

† Quoted by Bolt.

inner regions of the planet before they are picked up in places like Denmark and Greenland. The speed with which different kinds of wave move depends on the density of the material they are moving through, so their paths can be bent, refracted like light moving from air into glass, or reflected at boundaries between different layers of material. This affects their times of arrival at different places on the other side of the Earth. But it took a master of the black art to work out what the different arrival times of the waves at different locations was revealing.

The first person to identify the existence of a core inside the Earth was the British geologist Richard Oldham. He found that earthquake waves initially increase in speed as they move into the Earth, but that at a certain depth they suddenly slow down. He interpreted this as meaning that the waves had reached a liquid core within the Earth, below the solid rocks (mantle) on which we live, and published his results in 1906. What Lehmann discovered, and reported in 1936,* is that within this liquid core there is a denser, possibly solid core. She found that the data could be explained by the presence of an inner core about 2,800 kilometres across (1,400-kilometre radius), in which the waves travelled at about 10 kilometres per second:

> We take it that, as before, the earth consists of a core and a mantle, but that inside the core there is an inner core in which the velocity is larger than in the outer one … the inner core functions as a crude dispersing lens for seismic waves.

The radius of the Earth is 6,360 kilometres, and modern measurements give the radius of the inner core as 1,210 kilometres (a diameter of 2,420 kilometres) and the radius of the whole core as

* *Publications du Bureau Central Séismologique International*, volume 14, p. 87.

3,470 kilometres – pretty close to her figures. Her discovery was soon confirmed by other researchers, and she must have been particularly pleased when her former teacher Gutenberg published a paper in 1938, co-written with Charles Richter, confirming an inner core with a radius close to 1,200 kilometres and in which seismic waves travelled at about 11.2 kilometres per second. But it took decades for the precise details of what became known as the Lehmann discontinuity to be worked out, and for it to be confirmed that the inner core is solid.

During those decades, Lehmann did not rest on her laurels. She stuck to her day job until her 65th year, maintaining routine operations in Copenhagen even during the Second World War, and publishing 35 papers along the way (the first in 1926); but she became even more productive after she retired in 1953 and was free to pursue her own interests full time, although this later work is primarily of specialist interest. One of her key connections after 1953 was with the Lamont Geological Observatory in the United States, which she had first visited in 1952. She returned there many times, as well as visiting other seismological laboratories in the USA and Canada, including those in Berkeley and Ottawa. One of the first of these visits took her to the Seismological Laboratory of the California Institute of Technology for four months in 1954, where she worked once again with her old mentor Beno Gutenberg, by then the director of the laboratory. The decades following her retirement saw an upsurge in funding for seismological research, because of the need to monitor underground nuclear test explosions, and this led to the establishment of a global network of seismic stations (the Worldwide Standardized Seismographic Network, or WSSN), whose observations were freely published for all seismologists to see and use after 1960. This enabled researchers such as Lehmann to make more accurate measurements of earthquake waves than ever before, which she used to study the upper part of the mantle, the layer of hot rock beneath the crust and above the core.

In her seventies, Lehmann was instrumental in making the WSSN respectable to the scientific community. Nuclear testing was, and still is, a hot political topic, and seismologists might have been suspected of introducing political bias into their interpretations of the data from the WSSN. But Lehmann was, by the 1960s, of such high standing in the community, known for her integrity and ability to 'read' seismogram traces, that any results she endorsed could be taken at face value. Bruce Bolt, who knew her from 1960 until her death, has described her as a dedicated and thorough scientist with a highly critical and independent mind, who as a lone woman working in seismology before the Second World War needed strong resources of character to sustain her. Her great skill was to 'cut through the complications to the core* of the problem' in a variety of situations.

Among her many honours, Lehmann was elected as a foreign member of the Royal Society in 1969, and received honorary doctorates from the University of Copenhagen (in 1968, when she was 80, so not before time) and the University of Columbia. The Danish Geodetic Institute held a reception for her on her 100th birthday, in 1988, which was attended by geophysicists from Europe and America. By then, she was regarded as one of the most influential seismologists of the twentieth century.

But she never forgot her struggle to make her way in science in the early decades of the twentieth century. Lehmann died on 21 February 1993, a few weeks short of her 105th birthday. In a eulogy, her aunt's grandson, Nils Groes, recalled her telling him 'you should know how many incompetent men I had to compete with – in vain'.†

* We are not sure if the pun was intended!

† See Bolt.

CHAPTER 5

Irène Joliot-Curie (1897–1956)

Pioneered the development of artificial radioactivity

Our next subject started life with certain advantages compared with Lehmann, but still had to overcome many obstacles to achieve success. As her name suggests, Irène Curie, as she was born, did not have any difficulty persuading her parents to let her study science. But the way her mother encouraged her scientific aspirations highlights the failings of society in that regard. She is also worthy of note for what must surely be a unique achievement – she three times came within a hair's breadth of winning a Nobel Prize, twice with her husband and once with another colleague, but fortunately on a fourth occasion she did cross the line first.

Irène was born on 12 September 1897; three months later her parents, Marie and Pierre Curie, began the experiments that would result in their own joint Nobel award in 1903, for the identification of the radioactive elements polonium and radium. Although they hired a wet nurse for the baby and a part-time maid to help in their small apartment, a key role in Irène's upbringing was played by Pierre's father, Eugéne, a retired medical doctor, who moved in with them following the death of his wife when Irène was a few weeks old. She later said: 'My spirit had been formed in great part by my grandfather, Eugéne, and my reactions to political or religious questions came from him more than from my mother.'*

Eugéne's role was even more important after 1906, when Pierre was run down and killed by a horse-drawn wagon while crossing the street, leaving Marie not just with Irène but with her baby sister Eve, born on 6 December 1904. For a couple of years after Pierre died, Curie and a few of her colleagues, including the future Nobel Prize winner Jean Perrin and Pierre's former PhD student Paul Langevin, formed a cooperative to teach their own children outside the conventional educational system. Irène benefited enormously from this. She was by all accounts already a serious child who was not exactly unsociable, but was happy in her own company. When she received the Nobel Prize in 1935, her sister wrote an article about her in which she recalled that Irène had no interest in clothes or makeup, and when given a box of sweets she would eat one and put the rest away for later – incomprehensible behaviour to the younger girl! But their lives changed significantly as Irène entered her teenage years.

On 25 February 1910, Eugéne Curie died, and in the years that followed Irène became much closer to her mother. Later in 1910, as she turned thirteen, Irène started at a 'proper' school, the Collège de Sévigné, where she excelled at maths and science, to prepare her for university. The following months, however, brought both triumph and tribulation to the Curie family. In December, Marie, already a Nobel Prize winner and probably the most famous scientist in France, applied for election to the French Academy of Sciences, which had never had a female member. This was

* See S.B. McGrayne, *Nobel Prize women in science*, Secaucus, New Jersey, 1998.

a competitive election, with three candidates fighting for one place – the other two candidates were, of course, men. After a heated debate, on 23 January 1911 the academicians voted, and Marie came second, by one vote, to Edouard Branly, a physicist who worked on radio technology. Because the vote was so close, they took a second ballot. This time, Branly won by two votes. This prompted the members of the Institut de France, the umbrella organisation to which the Academy of Sciences belonged, to have its own vote on whether a woman would ever be allowed to join their ranks. The proposition was defeated by 90 votes to 52.

The next blow came a few weeks later, when the press got hold of the story that Marie had become the lover of Paul Langevin, a married man, and portrayed her in lurid terms as a scarlet woman, although he got off lightly in their pages. Marie and her daughters had to shelter for a time in a friend's apartment until the fuss died down, but it made the inevitable emotional impact on Irène and Eve.

On the scientific front, things went better for Marie, although the scandal did not go unnoticed. In the autumn, Marie was one of the participants at a conference on physics sponsored by the Belgian philanthropist Ernest Solvay – the first of a series of Solvay Conferences that would be attended by leading scientists over the following decades. She was the only woman there, in the company of men such as Albert Einstein and Ernest Rutherford, who didn't care what she got up to in her private life. But *l'affaire Langevin* was not ignored by the Swedish Academy of Sciences, who discussed it among themselves before deciding that, as the Nobel Prize was given for the work and not the person, she should receive a second award, this time in chemistry. Essentially, Marie got the two awards for the same piece of work – the physics prize, with Pierre, for identifying polonium and radium, and the chemistry prize, in 1911, for purifying samples of the two elements.

In December, fourteen-year-old Irène accompanied her mother to Stockholm to see Marie receive the award. But Marie was exhausted by hard work, the strain of *l'affaire Langevin*, and almost certainly by radiation sickness. She suffered a severe kidney infection, eventually requiring surgery, and had to spend long periods recuperating with friends or at a sanatorium. She was not able to resume her teaching duties at the Sorbonne until January 1913. The children were largely cared for by a governess during 1912, and Irène wrote many letters to her mother which show, among other things, her growing political awareness. On 28 July she wrote: 'I also saw that almost every day a British minister risks being killed by the British suffragettes. But it seems to me that it isn't a very clever way of proving they are capable of voting.'*

In the summer of 1913 Marie was well enough to take her daughters on a hiking holiday in the Alps with Einstein and his family. Hans, Einstein's older son, was the same age as Eve Curie, and the families got on well together. Einstein was impressed by Irène's intelligence and would become a lifelong friend of hers.

Back in Paris, by 1914 Marie was deeply involved in setting up the Radium Institute for research into radioactivity, and Irène, now sixteen, was preparing to enter the Sorbonne in the autumn. Everything seemed set fair when the girls went for a summer holiday in Brittany, in the care of their governess. But while they were there, Archduke Franz Ferdinand of Austria-Hungary was assassinated, setting in motion the chain of events that led to the First World War. After initial German successes, the western front stabilised, and by October it was safe for Irène and Eve to return to Paris, where Irène not only started her course in maths and physics at the Sorbonne, but enrolled on a nursing course with the intention of helping her

* Quoted by Brian.

mother, who had taken the initiative in setting up mobile X-ray units (using commandeered research instruments) to speed up the treatment of wounded soldiers at field hospitals. In a patriotic gesture, Marie also spent most of her Nobel Prize money buying war bonds, even though she knew that they would end up as worthless paper.

By November, Irène had enough training to accompany her mother on some of her expeditions, while continuing her studies at the Sorbonne; less than a year later, still only seventeen, she was competent enough to take temporary charge of the X-ray facilities at various front-line hospitals when Marie was busy elsewhere. On her eighteenth birthday, 12 September 1915, she was at Hoogstade in Belgium, where she spent the morning X-raying the hand of a soldier to locate bomb fragments, the afternoon watching a scratch football match, and the night sleeping in a tent. She still found time to complete her studies and be awarded her first degree in 1917, then be appointed officially as Marie's assistant at the Sorbonne. After the end of hostilities in 1918, Irène was awarded the Military Medal for her war work. Marie returned full time to establishing the Radium Institute as a leading research centre, now with Irène, working for her own PhD, as her partner. Both women, though, had received large doses of X-radiation as a result of using poorly shielded equipment.

Irène's research for her doctorate was supervised by Paul Langevin, who had remained a friend of the family after the end of *l'affaire*, and focused on the properties of alpha particles emitted by polonium – including, for example, their speed. But she took a break from her research in 1921 to travel with her mother and sister to the USA, to promote the work of the Radium Institute and thank the American people for the gift of a gram of radium (worth about $150,000 at 1921 prices) to be used in their research.* Marie was

* To put this in perspective, only about a kilo of radium has ever been purified. A little goes a long way in scientific research.

in poor health and easily exhausted, so Irène often gave speeches and received awards in her honour.

Back in Paris, Irène was sometimes referred to by her colleagues as the 'Princess' because of her privileged position as the daughter of the director of the Institute. She was described as having a quick temper (not suffering fools gladly), straightforward and direct, not pulling any punches in her debates about work. But there was no doubting her ability and her right to be working there. In 1922 a visitor commented that she was 'intimidating and robust, both intellectually and physically', but 'her imperturbability and her knowledge of physics and maths seemed well nigh incredible for someone age twenty-five'.* Off duty, however, with her few close friends and family, Irène showed a different side to her character, with a love of poetry (especially Robert Browning) and a fondness for dancing.

As the daughter of a double Nobel Prize winner, Irène's 'defence' of her PhD (the public lecture in which she described her work) drew the attention of the press and public in March 1925. A journalist from the Paris paper *Le Quotidien* asked her if a career in science might not be too demanding for a woman. 'Not at all,' she responded. 'I believe that men's and women's scientific aptitudes are exactly the same.' But what about family responsibilities, the reporter persisted. 'These are possible on condition they are accepted as additional burdens. For my part, I consider science to be the paramount interest in my life.' But by then, Irène had already met the man who would convince her that the additional burdens of family life would indeed be acceptable.

Jean Frédéric Joliot, always known as Fred, was born on 19 March 1900. At the time, he was the youngest of six siblings in a comfortably well-off family, but his father, Henri, had fought

* See McGrayne.

on the side of the workers in the Paris Commune of 1871. By the time Fred was old enough for high school, however, the family fortune had declined, and he had to attend a free school, the ESPCI (the École supérieure de physique et de chimie industrielles), where the focus was on vocational studies. This turned out to be a blessing in disguise. The director of studies at the school was Paul Langevin, who was so impressed by Fred (who graduated top of his class) that he recommended him to Marie Curie. After completing his compulsory military service, Fred started work at the Institute on 1 January 1925. While showing him round the place, Marie's deputy, André Debierne, mentioned that Joliot had chosen a bad time to start work there, because all the important discoveries in radioactivity had already been made. As if this were not off-putting enough, he was then handed over to the daunting figure of Irène, who had been given the job of bringing him up to speed with the work of the laboratory, especially with polonium.

Determined to catch up, Fred diligently followed her instructions and pestered her with questions. When there were too many questions to be answered during working hours, he took to walking her home to continue these informal seminars, and the two gradually discovered common interests. Their walks developed into weekend outings, where science was not the only topic of conversation. Fred later recalled:*

> I discovered this girl, whom other people regarded somewhat as a block of ice, an extraordinary person, sensitive and poetic, who in many things gave the impression of being a living replica of what her father had been. I had read much about Pierre Curie. I had heard teachers who knew him talking about him

* McGrayne.

and I rediscovered in his daughter the same purity, his good sense, his humility.

As for Fred, Eve Curie later described him as being in 1926 'the most brilliant and the most high-spirited of the workers at the Institute'.* He was the sort of person who, at a time when even close colleagues called each other 'Monsieur', 'Madame' or 'Mademoiselle', was universally known as Fred. He was charming and gregarious. It was to some extent an attraction of opposites – except intellectually.

One morning early in 1926, Irène casually informed Marie and Eve over breakfast that she was going to marry Fred. They were astonished, and Marie initially disapproved, suspecting that Fred was trying to further his career by riding on the coattails of the Curie name, and worrying about the age difference – he was nearly three years younger than Irène. Nevertheless, the wedding took place on the morning of 9 October 1926. In characteristic disregard for convention, after a lunch organised by Eve, the couple went back to work in the afternoon. That night, they each went to their separate apartments, beginning married life together the next day in Marie's apartment while she was away in Copenhagen. They chose to be known officially as Frédéric and Irène Joliot-Curie, but would use their original names on scientific papers. For the next few months, Marie remained cool towards Fred, and would introduce him to visitors at the Institute as 'the young man who married Irène'. This changed after Irène and Fred's daughter Hélène was born on 19 September 1927.† From then on, he became Fred to her as well as to everyone else.

* In her book *Madame Curie*.

† Hélène grew up to marry Michel Langevin, Paul Langevin's grandson.

Irène Joliot-Curie, 1921
Smithsonian Institution, via Wikimedia Commons

Typically, on the day Hélène was born Irène worked until lunchtime, then went off to have the baby and was back at work a few days later – although she had been diagnosed with tuberculosis and advised to take a long rest, and not to contemplate having another baby. Of course, she ignored both pieces of advice.

As a scientific team, Irène and Fred were as compatible as they were in marriage. Although her PhD was in physics, Irène concentrated on the chemistry, while Fred (soon to have a PhD in chemistry) concentrated on the physics. And they were blessed with the largest quantity of polonium available in any laboratory in the world to use in their experiments. Irène tended to think things out logically, step by step, while Fred tended to take a quick look at different possibilities before settling on a position. The only drawback was that neither of them had great experience in theory, so they did not always immediately see the best explanation for their experimental discoveries. Which is why they sometimes failed to take the final step on the road to Nobel glory.

The sequence of Nobel Prize near misses began in 1930, when two German researchers, Walther Bothe and Herbert Becker, reported that when beryllium was bombarded with alpha particles from polonium it gave off a powerful 'radiation' which could penetrate a sheet of lead two centimetres thick. Bothe suggested that this might be a more intense form of radiation similar to gamma rays – which are themselves essentially a more intense form of X-rays. Fred and Irène realised that this provided a new way to probe the structure of matter, and after replicating the German experiment they bombarded a thin sheet of paraffin with the new radiation. Paraffin was chosen because it contains a lot of hydrogen atoms, bound to carbon atoms, and sure enough the radiation knocked protons (hydrogen nuclei) out of the paraffin.

In January 1932, two months before their second child, Pierre, was born, they reported the discovery in a paper titled

'The Emission of Protons of High Velocity from Hydrogenous Materials Irradiated with Very Penetrating Gamma Rays'. But other physicists quickly realised that the effect could not be produced by gamma rays, which are a form of electromagnetic wave (like light) and could not carry enough energy to give the protons the required kick. Ernest Rutherford, head of the Cavendish Laboratory in England, accepted that the Joliot-Curies' experimental results were correct – he was sure they knew what they were doing – but doubted their interpretation. His number two at the Cavendish, James Chadwick, first repeated their experiment then went on to bombard a variety of other substances, including the elements carbon, nitrogen, lithium and beryllium, with the mystery radiation. Every time, protons were knocked out of the atoms. He concluded that the bombarding radiation must actually be a stream of particles, each with about the same mass as the proton, and gave them the name neutrons. As we mentioned in Chapter 2, further work showed that every atomic nucleus (except that of the simplest form of hydrogen) contains both protons and neutrons. Chadwick received the Nobel Prize in Physics for the discovery in 1935.

The second near miss came hot on the heels of the first. It involved a piece of apparatus called a cloud chamber, in which charged particles moving through air saturated with water vapour leave traces like the vapour trails of high-flying aircraft; these can be photographed to provide a permanent record of experimental results. The cloud chamber used by the Joliot-Curies had been built by Fred, who was a highly proficient engineer. In 1932 they decided to study the effect of neutrons, produced by bombarding beryllium and boron with alpha particles from polonium, on various substances. The tracks in their cloud chamber of the particles produced by this bombardment were photographed in the usual way. By applying a magnetic field across the cloud chamber, the charged particles could be made to follow curved trajectories, with positively charged protons curving one way and negatively charged electrons curving the other way. Because protons are much heavier than electrons, the two kinds of particles could also be distinguished because their tracks curve by different amounts. So far, so straightforward. But in one photograph there seemed to be an electron track curving the wrong way. When they published the picture showing this 'wrong way' electron in November 1932, they suggested that it must actually be an electron from outside the experiment moving the other way, towards their source. Four months later, the American Carl Anderson, who had been studying cosmic rays (charged particles from space) interacting with lead plates, independently reported that he had seen 'wrong way' electrons, but correctly explained these tracks as being due to positrons, a positively charged counterpart to the electron.* Anderson received the Nobel Prize in Physics for the discovery in 1936.

Meanwhile, Fred and Irène had continued their experiments, moving on to bombard aluminium with alpha particles and study the products of the interaction. To their surprise, the irradiated aluminium released both neutrons and positrons. Their initial conclusion was that the proton must be a compound particle, made of a neutron and a positron bound together, and that the aluminium nuclei were releasing protons which spontaneously turned into neutrons and positrons. In 1933, Fred and Irène were invited to participate in the latest Solvay Conference, where, in a sign of how little some things had progressed in the twenty-two years since the first such conference, out of forty-six scientists Irène was one of just three women, the others being her mother and Lise Meitner. Six of those scientists had already won Nobel Prizes, and a further

* Positrons are a form of antimatter, and can only exist briefly in our Universe before they meet a matter counterpart and the pair annihilate in a burst of energy.

twenty-four would soon receive this accolade. When Fred* presented their latest work to the meeting, it provoked a storm of objections from researchers who were convinced they had made a mistake, and had misinterpreted the experimental results. The main objection came from Meitner, who said that her team had carried out similar experiments and never detected this mixture of particles. Bohr, however, encouraged them to continue their experiments to find out who was right.

Chastened, but encouraged by Bohr, the Joliot-Curies began a new series of experiments late in 1933. In November, they received a letter from Meitner apologising and confirming that after further experiments she was now fairly convinced that positrons were being emitted in these interactions. So they were now in a race with the German team to get to the truth. On 11 January 1934, Fred was working alone in the lab, bombarding aluminium with alpha rays from polonium and monitoring the radiation from the aluminium targets with a Geiger counter, which clicked every time a fast-moving positron hit it. In order to test how the energy of the alpha rays affected the aluminium, he gradually moved the polonium source further away from the target, while monitoring the clicking of the counter. To his astonishment, he found that even when he removed the polonium entirely, the counter kept clicking, at a steadily declining rate. The target was still emitting positrons, even when it had stopped being bombarded. It had become radioactive in its own right. He called Irène down to the basement lab to listen to the sound of a new discovery.

The couple soon obtained similar results using boron and magnesium targets, and carried out chemical analyses to identify the radioactive materials being produced by the bombardment. It became clear that the alpha rays had not only caused one element

* Irène always insisted that he should present their joint work, to ensure that nobody thought he was just her assistant.

to be transmuted into another, but that the 'new' element was radioactive. In the case of aluminium, a nucleus of aluminium-27 plus an alpha particle (He-4) forms a radioactive nucleus of phosphorus-30 plus a neutron, and the phosphorus then emits a positron, converting one of its protons into a neutron, to become a stable nucleus of silicon-30. The chemistry involved in making the analyses had to be carried out with breathtaking speed and skill, since the radioactive elements produced in this way decay with half-lives of only a few minutes (which is why they are never found in nature). But it showed that the explanation of artificial radioactivity required an understanding of both the neutron and the positron, the two near misses of the Joliot-Curie partnership.

This time, there was no near miss. The discovery was reported to the French Academy on 15 January 1934, and quickly replicated in labs around the world. Fred and Irène received the Nobel Prize in Chemistry for their achievement in 1935 – the same year that Chadwick received the Physics Prize for the discovery of the neutron, and a year before Anderson received his for the discovery of the positron. The ailing Marie Curie lived to hear the news of the discovery, but died on 4 July 1934, at a sanatorium in the Alps. Years of exposure to radiation had finally caught up with her, as they would in due course with Fred and Irène. Indeed, by now Irène was also having to take long rest cures in the mountains while Fred held the fort at the lab and with the children. But there was still time for one more major contribution to science – involving another near miss.

The Nobel Prize changed the Joliot-Curies' lives both at home and at work. They had enough money to pay for a comfortable new house, with a study for each of them and a tennis court, and a new post of professor in nuclear chemistry was created for Fred at the Collège de France, the most prestigious seat of learning in the country. Irène found a new partner for her work at the Institute, a young Yugoslav called Paul Savitch. But she was no longer capable of the same intensity of work as before. Chronically ill, she was advised by her doctor to rest in the afternoons, and responded by having a cot brought in to her office at the Institute where she could take a nap.

As we mentioned in Chapter 2, in the mid-1930s Irène and Savitch found that when they bombarded uranium with neutrons they obtained more neutrons and a substance they could not identify, but which had similar properties to lanthanum, the element with atomic number 57. They called this R-3.5 because it had a half-life of 3.5 hours. Otto Hahn, in Germany, poured scorn on their claims,* and they backtracked, writing in a paper published in 1938 that 'this substance cannot be anything except a transuranic element' – that is, an element heavier than uranium. But as we have explained, the German team found that uranium is indeed broken apart by the neutron bombardment, producing barium, with atomic number 56. The rest, as they say, is history. But it is worth mentioning Lise Meitner's comments in a letter to Hahn chastising him for not giving enough credit to the French team. She pointed out that in the paper he had sneered at:

> They emphasised strongly that their 3.5 h substance had very remarkable chemical properties and emphasised the similarity to lanthanum… these findings led you to begin your experiments. And again you have not stated that quite clearly … Curie obviously saw that something remarkable was going on.†

* It has been suggested that he was deliberately trying to discourage the French team from following up this idea to leave him a clear run at the problem. That would be entirely in keeping with what we know about Hahn, but must remain speculation.

† Quoted by Gilmer, in M.-H. Chiu, Penny Gilmer, and D. F. Treagust. See also J. Vujic, http://www.k-grayengineeringeducation.com/blog/index.php/2009/02/11/engineering-education-blog-liese-meitner-and-nuclear-fission-2/

Once again, Irène had come close to gaining at least a share of a Nobel Prize, and it has been suggested that if she had still been working with Fred, who tended to make bigger intuitive leaps, they would have won this particular race.

Fred, however, was by then deeply involved in the construction of France's first cyclotron, a machine capable of accelerating charged particles to high velocities to allow, among other things, for the manufacture of many 'artificial' radioactive elements. These radioisotopes would become immensely valuable both in industry and medicine in the decades that followed.

Irène had become more politically active by the middle of the 1930s, and had reluctantly agreed to serve in government for a few months in 1936 as an under-secretary of state for scientific research, one of the first three women given ministerial posts in France, then to hand over to Jean Perrin by prior arrangement. She did this primarily to promote the feminist cause, not least by drawing attention to the fact that although she could be a government minister, she could not vote! French women were only given that right in 1945.* By then, of course, the world had once again been turned upside down by war.

Irène and Fred were both politically active on the left in the 1930s, openly supporting the anti-Franco forces in the Spanish Civil War, and were well aware of the implications of the rise of fascism. By the time war began,† Fred and his team were familiar with the possibility of using uranium fission to power a nuclear

* She also repeatedly applied for election to the French Academy, knowing that she would be rejected, in order to highlight their anti-women stance. This barrier only fell in 1962, with the election of Marguerite Perey, who had been one of Marie Curie's research assistants.

† Eve Curie, by then a journalist, left France early in 1940 and spent the next few years as a war correspondent.

reactor, and possibly a bomb. In a far-sighted move, before the German invasion of France he arranged for the French to purchase the entire stock of heavy water – a form of water in which hydrogen atoms in the molecules are replaced by the isotope deuterium, useful in controlling the speed of nuclear reactions – from the Norsk Hydro Company in Norway. In a subterfuge worthy of a James Bond movie,* all 185 kilos of the stuff, in aluminium containers, was flown from Norway to Scotland while a decoy aircraft headed towards Holland but was forced down in Hamburg, where a search revealed nothing interesting. This seriously hampered any efforts by the Nazis to build a nuclear reactor. By giving the appearance of cooperating (not collaborating) with the Nazis, Fred, who joined the (banned) communist party in 1942, was able to keep control of his laboratory, where his team carried out pure research on non-military topics. They also secretly manufactured explosives and radio equipment for the French Resistance. His work for the Resistance culminated in August 1944 when he participated in the rise against the occupying forces. He was awarded the Croix de Guerre and made a Commander of the French Legion of Honour, and in 1944 he was also elected to the Academy of Science.

By contrast, Irène, plagued by tuberculosis and other ailments, had a relatively quiet war, concentrating on looking after the family. She was allowed by the Nazis to travel to neutral Switzerland a couple of times for health reasons, leaving the children behind as, in effect, hostages for her return. But on 6 June 1944 they managed to cross the border with her, evading the German guards, who had been distracted by news of the Allied invasion of Normandy that day.

* It was indeed turned into a movie after the war, with Fred playing himself. The further adventures of the heavy water are worth looking up if you are interested in such things.

After the war, Fred became head of the French Atomic Energy Commission and oversaw the construction of the first French nuclear reactor, while Irène became a full professor at the Sorbonne. They remained active in both politics and science, but both became increasingly ill. Irène died of leukaemia on 17 March 1956, at the age of 58, and Fred just two years later, from cirrhosis of the liver caused by exposure to radiation. He was also just 58 when he died.

In an obituary of Irène in *Nature*, James Chadwick wrote:

> Her parents were both persons of strong and independent mind, and Mme. Joliot-Curie inherited much of their character as well as their scientific genius. She had a powerful personality, simple, direct and self-reliant. She knew her mind and spoke it, sometimes perhaps with devastating frankness; but her remarks were informed with such regard for scientific truth and with such conspicuous sincerity that they commanded the greatest respect in all circumstances. In all her work, whether in the laboratory, in discussion, or in committee, she set herself the highest standards and she was most conscientious in the fulfilment of any duties she undertook.

CHAPTER 6

Cecilia Payne-Gaposchkin (1900–79)

Discovered what stars are made of

Cecilia Helena Payne is the first of our female scientists to have been born after the start of 1900, on 10 May that year, at Wendover, some 40 miles northwest of London. From an early age, she learned that the education of boys took precedence over that of girls. Her father, a barrister, died when she was four (the same year her sister Leonora was born), and when she was twelve her mother moved the family to London to obtain a better education for Cecilia's younger brother Humfry, born in 1902, who went on to become an archaeologist. The move clearly worked for him, but not for Cecilia, who had to move from a school where she had thrived, progressing well in mathematics and science, to an old-fashioned Church of England school where there was initially no science tuition and even in mathematics she had to go back to the basics such as long division. Every year, all the pupils in the school took the same general knowledge examination. Cecilia, in her first year, came second in the whole school. Instead of praising her, the head teacher used this as an excuse to upbraid the older children for failing to beat her. In her autobiography* she wrote: 'It has taken me many decades to overcome the resentments and doubts engendered in a well-meaning but bigoted atmosphere.' Her response to the emphasis the school placed on religion was to carry out her own experiment on the power of prayer. Before one examination, she prayed for success; before another, she did not pray. She did slightly better in the second exam.

Things improved slightly after a new teacher encouraged Cecilia by lending her science books and taking her to museums, but initially her focus was on botany, regarded as a suitable thing for girls to study. In the war years, she tells us: 'I undertook the study of calculus and coordinate geometry by myself, and mathematics took on a kind of mystical significance.' But a new maths teacher, far from encouraging this enthusiasm, told her that she would never amount to anything, and she was told to leave the school in the year she turned eighteen. This was a key moment in her life.

St Paul's school, in Hammersmith, was persuaded to take Cecilia so that she could prepare for the entrance examination for Cambridge University. Although science was her first love, the music teacher, Gustav Holst, also urged her to take up a career in music, but science won. The principal of the school, Frances Gray, later recalled:

> Cecilia Payne had originality and ability far in advance of the originality and ability of the ordinary clever school-girl. I was very much struck, also, by her power of presenting a difficult subject in a lucid and attractive manner. I heard her give the Science Club of the School a remarkably clear and interesting lecture on Aviation. It would not have been unworthy of an experienced teacher of Physics.†

---

* See Bibliography. All quotes from her are from this source unless otherwise indicated.

† Quoted by Moore.

Almost from a standing start, in one year Cecilia worked so hard that in 1919 she won the Mary Ewart Scholarship for Natural Sciences, sufficient to fund her entire expenses to study at Newnham College in Cambridge, where she arrived in September 1919.

Initially, she studied the unusual combination of botany, physics and chemistry, with the intention of concentrating on botany after the first year. But her plans were changed by a lecture she attended on 2 December 1919. The speaker was the astronomer Arthur Eddington, and he had come to Cambridge to describe the results of an expedition to observe a solar eclipse that had confirmed the accuracy of Einstein's general theory of relativity. This had been front-page news in the papers less than a month before, and there was great demand for tickets. Newnham was given just four, none of them initially allocated to Cecilia. But when one of the ticketholders became ill, she passed hers on to Cecilia. The effect of Eddington's talk on her was shattering: 'For three nights, I think, I did not sleep. My world had been so shaken that I experienced something like a nervous breakdown.' From that moment, botany took a back seat, and at the end of her first year Cecilia switched her major to physics. This meant attending lectures at the Cavendish Laboratory, where she was usually the only woman in the audience. Among the arcane rules imposed on female students at that time,* she was required to sit in the front row. When she progressed to the advanced course in physics taught by Ernest Rutherford, the head of the Cavendish, he would begin every lecture by gazing intently at Cecilia and declaiming in a loud voice: 'Ladies and gentlemen …', producing a storm of noise as the male students stamped their feet in approval of his 'witticism'. But she was learning from the master – Rutherford was the man who had proposed the nuclear

* Which included the requirement to wear a hat when in town, and being forbidden to be alone with a man in a canoe.

model of the atom, with a dense central core surrounded by a cloud of electrons. And she also learned from the man who refined that model when Niels Bohr visited Cambridge to give a series of lectures explaining how the arrangement of electrons in the outer part of an atom produces the lines in the spectrum of light created by different elements. This would soon be the key that she would use to unlock the secret of the composition of the stars.

Although she had been drawn to astronomy by Eddington's talk, in her undergraduate years Cecilia could not switch to astronomy because the astronomers were not part of the physics department. This meant that before becoming an astronomer she learned a great deal of physics, making her ideally placed to be at the forefront of the merging of physics and astronomy that would shortly be explaining things like how stars work. And she found a way to be a physicist by day and an astronomer by night – Newnham College had its own observatory, a telescope housed in a dome in the college grounds.

Cecilia was taught how to operate the telescope by Leslie John Comrie (always known as L.J.), seven years older than her, who was working for a PhD in observational astronomy. The university also had a working observatory, a collection of telescopes just off the Madingley Road on the outskirts of town, where serious work was still possible in those days before the spread of light pollution. Eager to learn more, Cecilia attended an open night for the public at the Observatories, where she asked so many questions that the harassed assistant demonstrating the telescope went off to fetch the director, Arthur Eddington, to see this young woman. They returned to find her demonstrating the telescope to a young girl; when Cecilia became aware of Eddington's presence, she blurted out that she wanted to be an astronomer. He famously replied: 'I can see no insuperable objection.' She asked him to recommend some books for her to study – as he listed them, she told him she had already read them. Realising that her interest was serious, he gave her permission to use the library at the Observatories, opening up new avenues of exploration.

By the time she was completing her studies in 1923, however, there seemed no prospect of her finding a way to become a professional astronomer. She would not graduate, since, as we mentioned in Chapter 4, although they were allowed to study and take exams at Cambridge, women were not awarded degrees until 1948; what happened until then was that a successful candidate was awarded a certificate saying that they had passed the examinations and fulfilled the requirements that would, if they had been a man, entitle them to a BA, and if they had been a man, they would have got a BA. The first hint of an opportunity for Cecilia to escape from this Alice in Wonderland world came before she took these exams, in 1922, when she attended a lecture by another eminent astronomer.

The speaker on this occasion was Harlow Shapley, the director of the Harvard College Observatory, who was on a visit to the UK. The lecture was at a meeting of the British Astronomical Association (BAA), in London. L.J. Comrie, who was well aware both of Cecilia's ability and the lack of opportunities for a woman in astronomy in England, was a member of the BAA, and offered to take her as his guest and introduce her to Shapley. As she had with Eddington, Cecilia came straight to the point, telling Shapley that she wanted to go to the USA to work for him. The startled Shapley made an offhand response, and probably thought no more about it – until he received a letter from her in March 1923 reminding him of the encounter, reiterating her desire to work at Harvard, and adding that Eddington himself had 'advised me most strongly to come to Harvard if I could'.* Eddington provided a glowing ref-

* Letter in Harvard University Archives.

erence, and in a letter to Shapley, L.J. tried to assuage any doubts the great man might have about employing a woman:

> There is this to be said about her (between ourselves) – I believe she is the type of person who, given the opportunity, would devote her whole life to astronomy and that she would not want to run away after a few years training to get married.*

The upshot was that Shapley promised to find $500 for a research studentship to help with her support at Harvard for a year (later extended to two years). Even in 1923, this was hardly enough to live on, and Cecilia engaged in a concerted effort to find more funding, applying for every grant for which she might be eligible. Newnham came up with a one-year studentship worth £120, and she discovered that there was an obscure prize of £50 on offer for the best essay on the Greek text of one of the Gospels. Although she diligently compiled a suitably erudite essay, hardly surprisingly hers was the only entry. The £50 (the first money she ever earned) paid for 'outfitting myself for my journey to the New World'. On 10 September 1923 that journey began onboard RMS *Laconia*, at Southampton. Shortly before setting sail, in June 1923 she joined the Religious Society of Friends (the Quakers), and she remained a member for the rest of her life, although she does not seem to have attended many Quaker meetings. It is possible that this was another example of Eddington's influence, since he came from a Quaker family.

Waiting for her on the other side of the Atlantic was a treasure trove of astronomical data. The Harvard Observatory had a collection of hundreds of thousands of photographic plates, each

* Harvard Archive.

containing the spectra of many stars, obtained using the observatory's fifteen-inch* refracting telescope, which was the largest of its kind in the world when completed in 1847. All this data had been collected and collated by several women, known in those days as 'computers', but at the time Cecilia arrived nobody had been able to interpret all the spectra to determine the nature of the stars. As a research student, Cecilia was able to choose (with Shapley's approval) her own line of research, and she arrived with very definite ideas on what to do, with the aid of all those spectra:†

> What really inspired me was the announcement which I had heard before I left Cambridge, England, on the subject of the Adams Prize for the following year – the subject was the study of matter at high temperatures. ... And I said to myself, 'I will write a paper on the observational study of matter at high temperatures' ... I didn't expect to get the Adams Prize; I very much doubt whether a woman would be eligible. But I said to myself, 'At least I am going to make a contribution to the subject'.

The reason she felt able to tackle the problem was because in 1921 an Indian researcher, Meghnad Saha, had published a paper describing how the spectrum of an atom changes when it is ionised – when electrons are stripped away from its outer regions at high temperature. Each type of atom – each element – produces characteristic patterns of lines in the rainbow spectrum of light, as distinctive as a barcode. But the details of the barcode change when the atoms are ionised, making it harder to identify which atoms (elements) are present, and how much of each element is

* The size refers to the diameter of the main lens.

† Interview of Cecilia Payne-Gaposchkin by Owen Gingerich; see Bibliography

present – in, for example, the atmosphere of a star. The details also depend on the temperature, so Saha's equation offered a way to measure the surface temperatures of the stars, as well as to find out what they are made of. In 1923, it was clear that the atmosphere of the Sun definitely contained many, if not all, of the elements found on Earth, and it was generally assumed that if our planet were heated up to the temperature at the surface of the Sun (about 6,000 Kelvin), it would produce a spectrum like the solar spectrum. It was also assumed that other stars had much the same composition as the Sun. Up to a point, Cecilia's painstaking analysis of the Harvard spectra using Saha's theory confirmed these expectations. But there were two striking exceptions.

The task Cecilia set herself was monumental. Just saying 'she studied the spectra of the stars and found out what they are made of' barely hints at the effort involved. Those spectra were preserved as photographs on glass plates, obtained when light from the stars, after passing through the Harvard refractor, was spread out by a spectroscope to make a smeared-out patch on a glass plate treated with chemicals to make it sensitive to light. These smears of grey on the glass plates – a bit like the smear you might make with an inky thumb – were crossed by the lines in the spectrum produced by the presence of ionised atoms in the atmospheres of the stars being studied. Cecilia's knowledge of Bohr's quantum theory, learned from Bohr himself, gave her invaluable background for this work. Even so, it took months for her to learn to identify particular patterns of lines with the patterns Saha's theory predicted for particular elements in ionised states, and then to begin to use those measurements and Saha's theory to determine the temperatures of some of the stars. This led to her first publication on stellar spectra, and an important lesson from Shapley. He encouraged her to publish her results, but took her to task when she drafted the paper under the name C.H. Payne. 'Are you ashamed of being a woman?'

Cecilia Payne-Gaposchkin, Harvard College Observatory
Smithsonian Institution/Science Service, restored by Adam Cuerden, Public domain, via Wikimedia Commons

he asked. She changed it to Cecilia H. Payne, letting the scientific world know that the author was indeed a woman.

This decision not to hide her gender later played a big part in the life of another successful astrophysicist, Joan Feynman. Born in 1927, Joan wanted to have a career in science, but was told by everyone except her elder brother, Richard, that women could not be scientists. On her fourteenth birthday, 31 March 1941, Richard gave her a college-level book on astronomy and urged her to persevere with it. Ploughing through the text, on page 407 she came to an illustration showing part of the spectrum of a star. The caption credited the astronomer who had obtained the data – Cecilia Payne-Gaposchkin. It showed her that women could do science! As Joan recalled: 'The secret was out: it was possible! From that day on I was able to take my own interest in science seriously.'* If the caption had read 'C. Payne-Gaposchkin', the world might have lost another female scientist.

Once she had the knack of reading the stellar spectra, Cecilia produced a series of papers on the constitution and temperature of the stars, until Shapley suggested that she should put her work together as a thesis to be submitted for a PhD. She later recalled:†

> Shapley said, 'Why don't you get a doctor's degree?' I said, 'I don't want a doctor's degree. I have a degree from Cambridge. That is the highest degree in the world. I don't want any other.' He – well, I don't know, perhaps he stuck his neck out in getting me admitted as a student and wanted me to get a degree anyway. He persuaded me and I did.

* See Joan's contribution in *Most of the Good Stuff*, edited by Laurie Brown & John Rigden, American Institute of Physics, New York, 1993.

† Gingerich interview.

In fact, Shapley had an ulterior motive. He wanted to get astronomy recognised as an academic department at Harvard, and getting a PhD for one of his students was a step towards that goal. But he faced a big problem. The logical place for the thesis to be submitted was to the physics department at Harvard. But the chairman of that department, Theodore Lyman, refused to contemplate the idea of a woman receiving a Harvard PhD. Shapley found a compromise. He sweet-talked Lyman into agreeing to forward the thesis to nearby Radcliffe College, the female counterpart to Harvard, which gladly handled the formalities. Before the degree could be awarded, however, there was one more difficulty to overcome.

The major part of the thesis was both important and uncontroversial. She had shown that the atmospheres of all the stars analysed had much the same composition, and that the differences in the spectra from one star to another are caused by differences in temperature and pressure, not by differences in the composition of the stars. What's more, except for hydrogen and helium (the two lightest elements), the relative proportions of the different elements in the stars closely matched the abundances seen on Earth. But – a very big but! – hydrogen and helium did not fit the pattern. There seemed to be a thousand times more helium than expected in the atmospheres of stars, and a million times more hydrogen. Nevertheless, that was what the evidence implied, so that is what Payne put in her thesis. This discovery underpins some of the key concepts in modern astronomy and cosmology, from the understanding of how hydrogen and helium emerged from the Big Bang to the way heavier elements, including the atoms we are made of, are built up inside stars and scattered through space. But the astronomical establishment was not ready for this revolutionary idea in 1925.

Shapley sent the draft thesis to Henry Norris Russell, the director of Princeton University Observatory, to be reviewed before it was formally submitted to Radcliffe. Russell was impressed by everything in the thesis except for the conclusions about hydrogen and helium. He wrote to Payne to tell her that: 'I am convinced that there is something seriously wrong with the present theory. It is clearly impossible that hydrogen should be a million times more abundant than metals.'* Under pressure from Shapley, Payne rewrote that part of her thesis to say: 'Although hydrogen and helium are manifestly very abundant in stellar atmospheres, the actual values derived are … almost certainly not real.'

Payne got her PhD (from Radcliffe), the thesis was published as the first Harvard Observatory monograph, with the title 'Stellar Atmospheres', and Shapley soon got his astronomy department.† In 1929, using a different technique, Russell found that the stars are indeed mostly made of hydrogen, and in his published account mentioned that Payne had got there first. He noted that his results for all the elements, including the great abundance of hydrogen, agreed remarkably well with hers – 'the most important previous determination of the abundance of the elements by astrophysical means is that by Miss Payne'. But because he was an established senior astronomer and she was still an unknown junior researcher, for decades afterwards he received all the credit for the discovery.

She was certainly badly treated; but was she badly treated because she was a woman? In this case, almost certainly not. Russell would, it seems likely, have been equally dismissive if the thesis had come from a man. In her autobiography, Cecilia passed on to later generations the lesson she had learned: 'I was to blame

* Quoted by Moore. By the way, astronomers call everything except hydrogen and helium 'metals'!

† In 1962, the astronomer Otto Struve described her thesis as 'the most brilliant PhD thesis ever written in astronomy'. See O. Struve and V. Zebergs, *Astronomy of the 20th Century*, Macmillan, New York, 1962.

for not having pressed my point. I had given in to Authority when I believed I was right ... I note it here as a warning to the young. If you are sure of your facts, you should defend your position.' Even armed with a PhD, however, after 1925 Payne continued to work in the shadow of the male giants of astronomy of the time.

In the summer of 1925, Payne took her first holiday in two years, returning to England but discovering that it no longer felt like home. She now perceived the class structure of British society with the eyes of a visitor from the land of opportunity, and 'it supplied one of the strongest reasons for wishing to make [the USA] my home'.* On her return to Harvard, thanks to her thesis Payne was offered a job at the Lick Observatory in California, but instead she accepted a post as Shapley's assistant, with a tiny salary just sufficient to enable her to move out of the student dorm where she had been living and in to her own apartment. The salary was listed in the department accounts under 'equipment'.

As Shapley's paid assistant, she had to work at his direction on a project to measure the brightnesses of stars. This was so boring that she described the next couple of years as the worst of her life, as far as work was concerned. She also had to lecture on astronomy to graduate students, a role in which she excelled. But she was not listed as a lecturer or instructor, and not paid any more for this time-consuming work. It was not until 1930 that her salary was increased to $2,700 per year and she was officially allowed to devote the summer months to her own research, even though she was still listed as equipment. Harvard's president, Abbot Lawrence Lowell, told Shapley that 'Miss Payne would never have a position in the University' as long as he was alive. The increased freedom enabled her to take a road trip to California in a two-door Model

* Indeed, in 1931 she became a US citizen.

T Ford, visiting both the Mount Wilson Observatory and the Lick Observatory. But after that it was back to routine at Harvard, until her life was changed as a result of a visit to Europe in 1933.

The target of the trip was the Russian observatory at Pulkova (now Pulkovo), near what was then Leningrad and is now St Petersburg. But on the way she stopped off to visit astronomical centres in Cambridge, Leiden, Copenhagen and Stockholm. Any visit to Russia in those days was an adventure – the USA and Russia had no diplomatic contacts, Russia was in the grip of Stalin, food was scarce and the military presence was everywhere. She recalled:

> I spent two weeks at Pulkova, and felt I had experienced a lifetime. The atmosphere of tension never lifted. It was not only the drab and squalid living conditions of the man who was the Director of one of the great Observatories. It was not only the scarcity of food … Everyone was afraid – afraid to talk lest they should be overheard.

From Pulkova, Payne moved on to Göttingen, in Germany, to attend the annual meeting of the German Astronomical Society (Astronomische Gesellschaft), where she was approached by a young Russian exile (two years older than her), Sergei Gaposchkin. He had been on the losing side in the Russian Civil War that followed the revolution of 1917, but had escaped via Bulgaria to Germany. He had recently completed a PhD in astronomy and was working at an observatory near Berlin. But he was a Russian by birth, an illegal immigrant with no official papers, and the new Nazi regime had told the observatory to get rid of him. The next step would probably be a concentration camp. He needed help to get out of the country.

His story opened Payne's eyes to what was going on in Germany, and she had seen for herself how bad things were in Russia; she resolved to help him: 'Perhaps this, I thought, is my one chance to do something for someone who needs and deserves it.' Back in the USA, thanks to her intense lobbying on his behalf, Gaposchkin was offered a post as research assistant by Shapley; this offer was the key that made his presence in the country acceptable to the American authorities, who granted him a visa, and he arrived in Boston on 25 November 1933. In March 1934, to the astonishment of everyone who knew them, Sergei and Cecilia were married at New York City Hall. For a honeymoon, the couple travelled west to visit the great observatories, and returned home by ship via the Panama Canal. From now on, she would be known as Cecilia Payne-Gaposchkin. But apart from adding Sergei's surname to hers, she did not conform to the behaviour expected of a married woman.

Their first child, Edward, was born in May 1935, which didn't stop Cecilia from presenting a paper to the American Astronomical Society just four months before giving birth. Katherine followed in 1937, and, in a reversal of the expected role models, a few months after the birth Sergei took care of the baby for six weeks while she went to the Yerkes Observatory to give a series of lectures. Their third child, Peter, was born in 1940, again without interrupting the flow of her work.

Much of that work, especially on variable stars, was carried out with Sergei, but there was never any doubt who was the leader of the team. Sergei was a competent astronomer, but always in the shadow, academically, of his brilliant wife. She also studied the nature of bright stars ('stars of high luminosity') both in our Milky Way galaxy and in its neighbouring star systems, the Magellanic Clouds. These studies, which led to a stream of scientific papers and books, laid much of the groundwork for understanding how stars change as they age – stellar evolution. This was all important work, although nothing as dramatic as the discovery of what stars are made of. But while her reputation grew, her position at Harvard was still held back by her gender. Her salary remained stuck at $2,700 (Sergei got even less), and it was only in 1938 that her job description was changed to 'Astronomer', then to 'Phillips Astronomer' (in recognition of an endowment to the university), thus making her officially an 'officer of the university'. Lowell, who had vowed that she would never hold a position at the university, had retired in 1933, but lived until 1943, long enough to see this much of her progress. But the university only agreed to this on the explicit understanding that the appointment was not the equivalent of a professorship. Nevertheless, after the passage of another seven years, in 1945, for the first time a student perusing the university's catalogue to decide which courses to follow could find that there was an introductory course in astronomy taught by 'Cecilia Payne-Gaposchkin, Astronomer'.

In 1954, Shapley retired and Donald Menzel took over as director of the Harvard College Observatory. Astonished to discover how little she was paid, he doubled Cecilia's salary, and soon set the wheels in motion for a historic, long-overdue step. In June 1956 she was appointed as professor of astronomy, the first woman to be promoted to a full professorship at Harvard. The news was sufficiently significant to be recorded in the pages of *The New York Times*. At a party to which she invited all the female astronomy students at Harvard to celebrate the occasion, Cecilia, who was almost six feet tall, quipped that she found herself 'cast in the unlikely role of a thin wedge'.*

She soon became chair of the astronomy department at Harvard – the first woman to head a department at the university – and administration and teaching left her little time for

* Quoted by Virginia Trimble in Haramundanis.

further research. She retired in 1966, and died of lung cancer on 7 December 1979, a victim of the smoking habit she had begun in her teens. Sergei outlived her by nearly five years and died on 17 October 1984.

Cecilia Payne-Gaposchkin left two pieces of advice for anyone tempted to follow in her footsteps. When she received the Henry Norris Russell Prize* in 1976, to 'commemorate a lifetime of pre-eminence in astronomical research', she said:

> The reward of the young scientist is the emotional thrill of being the first person in the history of the world to see something or understand something. Nothing can compare with that experience … The reward of the old scientist is the sense of having seen a vague sketch grow into a masterly landscape.

And in her autobiography she advised:

> Do not undertake a scientific career in quest of fame or money. There are easier and better ways to reach them. Undertake it only if nothing else will satisfy you; for nothing else is probably what you will receive. Your reward will be the widening of the horizon as you climb. And if you achieve that reward you will ask no other.

That advice could almost have been written for our next subject, Cecilia's near-contemporary Barbara McClintock.

* The irony is delicious!

CHAPTER 7

Barbara McClintock (16 June 1902–2 September 1992)

Discovered how genes turn on and off

To some extent, Barbara McClintock had an easier passage into a scientific career than some of the other subjects of this book, because she studied plants, and in the first decades of the twentieth century that seemed a much more ladylike occupation than physics or chemistry. But she still found her path to promotion blocked by a glass ceiling and had to wait a very long time for the importance of her work to be recognised; the pattern of her life followed almost exactly the template for successful female scientists in the early twentieth century identified by Bente Rosenbeck and described in the introduction to this book.

Barbara was the third child of Thomas and Sara McClintock, who had married just before Thomas graduated from Boston University Medical School in 1898. They settled in Hartford, Connecticut. Although Sara had a small inheritance which she used to pay off his university debts, the arrival in quick succession of three daughters (Marjorie in 1898, Mignon in 1900 and Barbara on 16 June 1902) and a son (Malcolm, but always known as Tom, early in 1904) put a financial strain on the family, and a physical strain on Sara. As a result, at the age of three Barbara* was sent to live with an aunt and uncle in Massachusetts, where she spent most of her time until she started school.

Three years later, the family finances had improved as Dr McClintock's practice grew, and the family, including Barbara, moved to Brooklyn (which was still largely undeveloped). She seems to have been an independent child, happy in her own company, and something of a tomboy who enjoyed games such as baseball and basketball, climbing trees and 'the same kind of free time [as] my brother'.† In her teenage years, she also displayed what was regarded as an unladylike interest in science and a passion for understanding how the world worked.

The First World War nearly nipped that passion in the bud. Her father was away as an army medic, and Sara exerted a considerable influence on the girls at a crucial time in their lives. Marjorie and Mignon were both bright students – Marjorie was offered a scholarship at Vassar – but Sara was convinced that too much education would make a woman undesirable as a wife, and regarded marriage as the only proper 'career' for a woman. Both the older girls did indeed marry and settle down. Barbara resisted the pressure to conform, in spite of her mother's fears that her daughter might become a college professor, a fate she seemingly thought worse than death. The situation was resolved when Dr McClintock returned from overseas and agreed to let her go on to university. So, in 1919, Barbara enrolled at Cornell University, in the College of Agriculture: 'I was entranced at the

---

* She was originally named Eleanor, but when she was four months old her parents decided that the name didn't suit her, and legally changed it to Barbara.

† Quotes by McClintock, unless otherwise specified, are from the extensive interviews conducted by Evelyn Fox Keller for her biography.

very first lecture I went to. It was zoology, and I was just completely entranced.'

Cornell was unusual at that time in having been founded on the principle of admittance for any person, and had accepted its first female student in 1872. It hadn't quite achieved an equal balance of the sexes by 1919, but in the year Barbara got her BSc, 1923, a quarter of the graduates from the College of Agriculture were women. Unfortunately, as we shall see, the situation was not so good at faculty level.

In her first two years at Cornell, Barbara joined in the usual student activities, and even joined a sorority, but she quit when she realised how snobbish it was. Boys featured strongly, until she decided that she was not the sort of person who was cut out for long-term relationships; she later told Evelyn Fox Keller, her biographer, that she never understood the point of marriage. Focusing on study from then on, she avoided any time-wasting activities, even having her long hair cut short, a couple of years before the shingled look became fashionable, because it was too much bother to look after. For a time, she played banjo in a jazz group, but even gave that up when she found that late nights playing in clubs interfered with her studies.

Studying was not work to her; she found delight in learning, pleasure in finding things out, which led her to excel. Halfway through her third (junior) year she was invited to take the graduate course in genetics, and she became a graduate student in all but name, although she could not officially graduate until the end of the following academic year. It was at this point that a problem about her gender arose for the first time. Genetics was taught in the plant breeding department, and they refused to accept a female graduate student. Instead, she had to register in the botany department; a trivial difference at one level, but a sign of things to come.

The work that McClintock carried out as a graduate student set the tone for her career. It involved studying the individual chromosomes, the carriers of genetic information, inside the cells of maize plants, and relating these to the characteristics of the plants to find out how differences in the chromosomes affect the appearance of the whole plant. At that time, nobody knew that chromosomes are made of DNA, and that DNA carries the instructions that determine the nature of the organism, but since she was dealing with whole chromosomes, that didn't matter. She developed techniques to tell chromosomes apart under the microscope, from their size and shape; maize is a particularly good material for this kind of study because it has large chromosomes and very visible differences that can be seen in the patterns of colour in the seeds* and other properties. The snag is that to follow changes from generation to generation takes years of study, and by the 1920s many geneticists preferred to work with the fruit fly drosophila, which reproduces much more quickly (a new generation every ten days) but has the disadvantage that the chromosomes are much smaller and harder to identify. Either way, the study of how the visible structure of chromosomes is related to the genetics of organisms and heredity is called cytogenetics.

McClintock finished her graduate studies and received her PhD in 1927, picking up an MA along the way, then stayed on at Cornell as an instructor (a teaching position with no guarantee of long-term employment) to study the way certain sets of genes called linkage groups, which are passed on together from one generation to the next, are physically associated with particular chromosomes. There were two ways of tackling such problems at that time. One involved breeding plants and studying the changes over the generations,

* Wild maize produces cobs with seeds of many colours, not the uniform yellow of supermarket sweetcorn.

the other studying the chromosomes themselves. But McClintock wanted to do everything. Since this involved a lot of literal field work, with her usual love of the practical and disdain for convention she usually dressed in a version of the kind of trousers known as plus fours.

Although she worked alone on her own project, she was part of a small group of cytogeneticists who carried out pioneering work at Cornell in the late 1920s and early 1930s. The most important of these colleagues were Marcus Rhoades and George Beadle (both born in 1903), who were then graduate students and were heavily influenced by McClintock. Their careers would follow the conventional path through academia to become distinguished professors, but although they both recognised that McClintock was something special, out of their league, there was no prospect of such a career for her. She worked at what she wanted to do, well aware that an instructorship was about all that she could expect. The conventional fate of female scientists at that time is highlighted by the example of a young woman who joined the Cornell team in 1929 – a fate strikingly different from those of her near-contemporaries Rhoades and Beadle.

Harriet Creighton was twenty years old when she arrived in Cornell with a fresh degree from Wellesley College. McClintock took her under her wing, acting as her thesis advisor and giving her a project to work on. Although Creighton didn't realise it at first, this was not some trivial piece of work for a student to cut her teeth on, but something at the cutting edge of current research. By 1929, geneticists had an understanding that the way properties (or traits) of an organism can be inherited must involve a physical 'crossing over' of bits of one chromosome to another chromosome in the process by which sex cells are made (meiosis). Each cell starts out with two sets of chromosomes, one from each parent, but in meiosis they get cut up and rejoined so that a single chromosome that is passed on to the next generation can contain bits (of, we now know, DNA) coming from each parent. But though the theory seemed sound, nobody had observed the process under the microscope. Observing chromosomes being shuffled in this way would be the last link in the chain establishing the basis of genetics, and that is what Creighton and McClintock achieved, with results published in the *Proceedings of the National Academy of Sciences* in August 1931. The following year, their work was one of the highlights of the Sixth International Congress of Genetics, held in Ithaca, New York.

This was the pinnacle of Creighton's career. Had she been a man, she would surely have been offered a research position somewhere after completing her PhD in 1933. As it was, she stayed at Cornell as an instructor for another year, then bowed to the inevitable and took up a teaching position at Connecticut College. Although she did later return to Wellesley as a faculty member in the botany department, she made no more contributions to cytogenetics.

By the time of the Sixth International Congress, even McClintock had realised the futility of hanging on at Cornell, although she was gaining recognition elsewhere. She knew that a proper faculty position was out of the question, and, she told Keller, clinging on as an instructor would have been embarrassing both for her and for the university. A temporary solution came in the form of a two-year fellowship that she was offered by the National Research Council (NRC). This enabled her to spend the years from 1931 to 1933 as a kind of nomadic scientist, criss-crossing the country in a Model A Ford, visiting the University of Missouri, the California Institute of Technology (Caltech) and Cornell, where she had to return regularly to check her crops – they couldn't give her a job, but at least they helped her in practical ways.

It was during this period that McClintock made her next major contribution to cytogenetics. The plant breeders were baffled by plants that seemed to be losing small bits of their genetic material, so that their offspring were variegated in unpredictable ways. She suggested that in crossing over, pieces of chromosome might get accidentally snipped off by the cell's mechanism, but instead of floating around inside the cell as a short length of material which could be used by the cell, it formed a loop, with the ends joined together. Such a ring chromosome would in effect be invisible to the cell mechanism, because it would not be possible for the cell to 'read' the information it contained. McClintock devised the mechanism and gave these fragments the name 'ring chromosomes' before anybody saw them, but following her lead the Caltech researchers did indeed identify these rings.

In 1933, after her NRC grant ran out, McClintock received a Guggenheim Fellowship to visit Berlin to work with German geneticists at the Kaiser Wilhelm Institute. This was the same year that Cecilia Payne visited Germany, and like Payne, McClintock was unprepared for the political reality there and was deeply affected by it. Hardly surprisingly, she achieved nothing in the way of research, and returned to Cornell in December with no job and no prospects, although she was still allowed to breed her corn there. It was at about this time that she suddenly realised:

> this was what they called a career for women … At this stage, in the mid-thirties, a career for women did not receive much approbation. You were stigmatizing yourself by being a spinster and a career woman, especially in science … I had gotten myself into this position without recognizing that it was where I was going.

Another temporary solution was found thanks to the efforts of Rollins Emerson, who was chairman of the plant breeding department at Cornell and dean of the graduate school. Even with those credentials, he could not get McClintock a faculty post, but with the aid of Thomas Hunt Morgan, whose work with drosophila had earned him the Nobel Prize in 1933, he was able to persuade the Rockefeller Foundation to fund her research for a year, notionally as Emerson's assistant, although this was a complete fiction. She received $1,800 dollars for the year, starting on 1 October 1934, and this was later renewed for a second year on the firm understanding that that would be the end of their support. But these were not productive years for her research, and her obvious resentment of the contrast between the way she was treated and the careers of younger male colleagues such as Rhoades gained her a reputation as a troublemaker with a chip on her shoulder. A better term is the one she used to describe herself: 'there was no hope for a maverick like me to ever be at a university'.

During the brief time that she did work at the University of Missouri, in Columbia, one Sunday she came in to work in her lab but found that the building was locked and she had forgotten her keys. The natural thing to do, in her mind, was to climb in through a window. So she did, much to the amusement of passers-by. She also ignored rules that seemed pointless. Officially, graduate students were not allowed to stay in the lab after 11pm, but she encouraged them to stay on if they were in the middle of an important piece of work. This kind of behaviour was by no means unique, and the kind of thing that was usually treated as harmless eccentricity in the case of a male academic (Richard Feynman is an obvious example). But to be both a woman and a maverick was to be beyond the pale. The authorities began to regard her as a troublemaker.

Her position at the University of Missouri was in any case precarious. It stemmed from the work she had carried out there as a visitor, invited by the geneticist Lewis Stadler. When he received a major grant from the Rockefeller Foundation to establish a centre for genetics in Columbia he was able to persuade the university

Barbara McClintock, 1947

Smithsonian Institution/Science Service; Restored by Adam Cuerden, Public domain, via Wikimedia Commons

authorities to offer her a post as assistant professor in 1936.* It was untenured, on a salary of only $3,000, on the bottom rung of the academic ladder with no hope of promotion even to the rank of associate professor; but it provided her with a base to work (although her corn was still grown at Cornell). She carried out important work there, using X-rays to break chromosomes into pieces and study the way they joined back together, providing insight both into the way mutations occur and the mechanisms of the cell which only allow chromosomes to rejoin in specific ways, not at random. Outside Columbia, her reputation grew, and in 1939 she was elected as the vice president of the Genetics Society of America. But nothing seemed to enhance her status at the university. In the summer of 1941, she decided to clear the air by asking the dean what prospects there were for her future. He replied that 'if anything happened to Stadler, they would probably have to fire me'. She immediately took an unpaid leave of absence, and left Columbia. Two months later, in August 1941, the dean heard a rumour that McClintock had been nominated for membership of the National Academy of Sciences. He contacted her offering a promotion and a pay rise if she returned, but she'd had enough of the University of Missouri and declined. The 'leave of absence' turned into a resignation. She had just passed her 39th birthday and, once again, she was unemployed.

Meanwhile, following a conventional (male) career path, her old friend and colleague Marcus Rhoades, a year younger than her, had become an associate professor at Columbia University in 1940. She wrote asking him where he was going to grow his corn, and

* Don't be fooled by the title. In American universities everyone is a professor. Assistant professor is the lowest rank, then associate professor, and finally professor. The UK equivalent of the post McClintock was offered would be junior lecturer.

he said he was thinking of using the facilities at a lab in a hamlet known as Cold Spring Harbor,* on the northern shore of Long Island. On impulse, she decided to go there and check out the possibilities. The geneticists there welcomed her with the proverbial open arms as a summer visitor, and when the summer was over she stayed on in New York, until in December she received an offer she couldn't refuse.

McClintock's main contact at Cold Spring Harbor was a geneticist called Milislav Demerec, who worked with drosophila. He had arranged the summer visit, and on 1 December 1941 he was appointed as director of the department of genetics at the Cold Spring Harbor lab. He immediately offered McClintock a post there for a year, which she accepted for lack of anything else, then early in 1942 he suggested turning this into a permanent appointment. Again, she accepted with some reluctance, but it proved to be the ideal long-term home for her.

At that time the department of genetics at Cold Spring Harbor was funded by the Carnegie Institution of Washington, and was outside the university system. McClintock was now free to work on whatever she wanted, with no teaching or administrative duties, free from bureaucratic rules about hours of work. It was almost like being a gentleman amateur scientist of centuries gone by, or a latter-day Eunice Foote. Much of the lab's work over the next few years focused on the war effort – including a major role in the development of the manufacture of penicillin – but McClintock continued with her studies of what happens to chromosomes when they are broken apart and rejoin.

1944 was a particularly good year for McClintock. First, she was invited by George Beadle, now a professor of biology at Stanford,

* Named, logically enough, after the cold freshwater springs that flow in the area.

to visit his lab to tackle the problem of identifying the chromosomes in a mould called Neurospora, which he was studying in his genetic research. Before she could make the trip, she really was elected (no rumour this time) as a member of the National Academy of Sciences. This is the highest professional honour for an American scientist, and she was only the third woman to be elected. She wrote to a colleague who congratulated her: 'I must admit I was stunned. Jews, women and Negroes are accustomed to discrimination and don't expect much.' Her work on Neurospora was a triumph, resulting in a complete 'map' of its chromosomes, which she could trace through the entire process of meiosis. And then, still in the same year, she became the first woman to be elected as president of the Genetics Society of America. This looked like the pinnacle of her career. Astonishingly, though, at the age of 42 her most important work was still to come, and its significance would not be widely recognised for decades.

One reason for her work being overlooked for so long is that it was also in 1944 that Oswald Avery published news of the discovery of the role of DNA in genetics, and in the decades that followed the attention of geneticists focused on the details of the role of DNA in heredity. Instead of studying complex organisms such as maize or drosophila, the experimental subjects of choice in these studies were bacteria. Bacteria reproduce in a matter of minutes, providing many generations to study, and they have a simple internal structure which makes it easier to analyse the structure of their genes. McClintock seemed to represent a hangover from the past, breeding plants and developing an understanding – a feel – for the entire organism. She almost literally spoke a different language from the new wave of molecular biologists, many of them with backgrounds in physics, not biology, who did not appreciate what she was doing.

What she was doing took years, but we shall describe the significance briefly. Most maize plants have green leaves, but some are yellow, or a lighter shade of green. But on individual leaves there may sometimes be odd markings – a streak of pale colour on a dark leaf, or a yellow blob on a green leaf, and so on. McClintock knew that each complete leaf developed from a single cell at the stem of the plant, which divides repeatedly to make the whole leaf. All the cells in the leaf ought to contain the same chromosomes – the same genes. So the differences must be caused by a copying error, a mutation, in the single cell that divides at the start of the streak of different colour. She could identify the exact cell in which the mutation had occurred, study its chromosomes and those of its daughters, and track the changes that resulted.

There were other oddities, affecting things such as the colour of the seeds in a corn cob and the frequency with which mutations occurred in different plants. McClintock was able to explain all this by showing that the genes which determine the overall features of an organism are not all 'on' together, but can be turned on and off by other genes, called control genes. One kind of control gene sits on the same chromosome as the gene that it controls, and it can 'switch' it from green to yellow, or whatever. Another kind, dubbed a 'suppressor' gene, effectively turns other genes permanently off, while yet another kind of control gene, which she called a regulator, can be anywhere in the cell, on any of the chromosomes, and can speed up or slow down the rate at which changes occur. And these control genes were able to move from one chromosome to another, a process called transposition, so that they might influence different parts of the genetic package, instead of being fixed in place, always giving the same instructions to the same genes. The control genes are actually copied by the cell's mechanisms and the copies inserted into different places. They do not physically move from one place to another. But they have become known as 'jumping genes', or, more formally, transposons. This means that the genetic package as a whole is not fixed and unchanging in the form inherited from the parent(s), but adapts itself in some ways as an organism develops, responding to outside influences.

All this had been worked out and established by McClintock by the end of the 1940s. In 1950, she published a paper presenting these discoveries in the *Proceedings of the National Academy of Sciences*, and a year later reported everything at a major symposium at Cold Spring Harbor, attended by most of the leading geneticists. At the end of the presentation, nobody made any comments or asked any questions. The audience simply had not understood, let alone appreciated, the significance of her work. After a few more attempts to get it noticed at scientific meetings, after 1953 she gave up pushing the idea, and only published her further results in the annual reports of the work at Cold Spring Harbor.

There things rested until 1961, when the French researchers Jacques Monod and François Jacob published a paper in the journal *Comptes Rendus* in which they announced the discovery of control genes at work in E. coli bacteria. Their paper made no mention of McClintock's work because they knew nothing about it. And even though she promptly published a paper which drew attention to the links with her work, the connection was still not properly appreciated, because molecular biologists were so focused on the behaviour of genes (by then known to be made of DNA) in simple bacteria that they had no understanding of more complex organisms, such as plants and people, with their larger and more complicated burden of chromosomes.

From the early 1960s onwards, as the molecular biologists developed new techniques and improved their understanding of genes, the significance of McClintock's work gradually became clear to more people. At the same time, biologists eventually realised that transposons were not just peculiar to maize but were widespread across species. At its simplest, you could say that the molecular biologists were working from the bottom up – strands of DNA – to the complications of chromosomes, while McClintock had worked from the top down – whole organisms— to the behaviour of chromosomes. Eventually, the two approaches met.

Meanwhile, McClintock received a string of honours. In 1965 she was made an Andrew White Professor-at-Large by Cornell, in 1967 she was awarded the prestigious Kimber Genetics Award by the National Academy of Sciences, and in 1970 she received America's top scientific award, the National Medal of Science, presented by the President – the first woman to receive this honour. By the mid-1970s geneticists were using the term 'transposable elements' to refer to pieces of DNA that could be inserted into different places in the genome (the whole package of chromosomes), with explicit acknowledgement of McClintock's work, and it was beginning to be understood that transposons must play a part in the development of a complex organism such as a maize plant or a human being from a single cell. Obviously, not all the genes in every cell in your body, for example, are turned on at once. Complex multicellular organisms like us (or maize) are made of cells each with identical genomes, even though the cells have different functions. A liver cell only 'uses' the genes that make it a liver cell, even though its chromosomes carry the entire recipe for making and running every kind of cell in a human body. Control genes are the key to this process.

By the 1980s, it was clear that the genomes of complex organisms are not fixed, but are capable of re-arrangement and change, which offers new insights into both development and evolution – and into cancer, caused when these mechanisms go wrong. In a paper published in 1980, McClintock said 'there is little doubt that genomes of some if not all organisms are fragile and that drastic changes may occur at rapid rates [which] could give rise to new species'. Jumping ahead a little, with the modern ability biologists have to measure the entire sequence of DNA in a genome, we now know that roughly two-thirds of the material in the human genome is actually in the form of transposons. People are so complicated that there is more DNA controlling what gets turned on and off than there is being turned on and off!

Our favourite example of the flexibility of the genome comes from the study of human twins. Identical twins develop from a single fertilised egg that divides into two. So they start out with identical genomes. But even though they may look alike as adults, inside there can be significant differences. If, for example, one twin develops type 1 diabetes, which is an inherited disease, the other twin usually will not. In one twin, the genes responsible for diabetes have been turned on, and in the other they have not. We don't know how or why this could happen, but it does; and clearly control genes are responsible.

In the 1980s, just past her own 80th birthday, McClintock was positively showered with awards and honours, including the Nobel Prize for Physiology or Medicine, 'for the discovery of mobile genetic elements', in 1983, when she was 81. She was the first woman to receive the Prize in this category unshared. In her Nobel Lecture, she said.

> Because I became actively involved in the subject of genetics only twenty-one years after the rediscovery, in 1900, of Mendel's principles of heredity, and at a stage when acceptance of these principles was not general among biologists, I have had the pleasure of witnessing and experiencing the excitement created by revolutionary changes in genetic concepts that have occurred over the past sixty-odd years. I believe we are again experiencing such a revolution. It is altering our concepts of the genome: its component parts, their organizations, mobilities, and their modes of operation. Also, we are now better able to integrate activities of nuclear genomes with those of other components

> of a cell. Unquestionably, we will emerge from this revolutionary period with modified views of components of cells and how they operate, but only, however, to await the emergence of the next revolutionary phase that again will bring startling changes in concepts.

McClintock had officially retired in 1967 but was made a Distinguished Service Member of the Carnegie Institution of Washington, an honorary post which enabled her to carry on working at Cold Spring Harbor, often with graduate students, where she was a regular visitor for the rest of her life, frequently giving talks on the history of genetics research and her part in it for young researchers. Late in life, she said: 'I just have been so interested in what I was doing and it's been such a pleasure, such a deep pleasure, that I never thought of stopping … I've had a very, very satisfying and interesting life.'* She died in New York on 2 September 1992, at the age of 90.

On 4 May 2005, the US Postal Service issued the 'American Scientists' commemorative postage stamp series, a set of four 37-cent stamps depicting Barbara McClintock, John von Neumann, Josiah Willard Gibbs and Richard Feynman; impressive company, in which she definitely belongs.

* See S.B. McGrayne, in *Nobel Prize Women in Science*, Carol Publishing Group, Secaucus, NJ, 2001.

CHAPTER 8

Dorothy Crowfoot Hodgkin (12 May 1910–29 July 1994)*

Used X-rays to determine the structure of vitamin B_{12}

By the end of the 1920s, there was the beginning of a recognised career path in Britain for a young woman who wanted to become a scientist. If she had the right background, there was the possibility of studying for a university degree. And if she was exceptionally gifted and hard-working, she might progress on to research. Our next subject might seem to have had it easy compared with her predecessors, but there were still obstacles to overcome. She was, for example, the first woman at the University of Oxford to be granted paid maternity leave (indeed, the first person at an Oxford college to give birth while holding a fellowship), and for most of her life she suffered from a severe physical disability. She is also, at the time of writing, the only female British Nobel Prize winner in science, which alone makes her worthy of inclusion.

* This chapter draws on Georgina Ferry's definitive biography of Hodgkin.

Dorothy Crowfoot,* as she began life, was born on 12 May 1910 in Cairo. Egypt was then under British control, and her father, John, worked as a senior civil servant in the department of education. Her mother, Mary (known as Molly), was one of the daughters of the possessor of a small country estate in Lincolnshire, sufficiently affluent to own a villa on the Italian Riviera. They had married in July 1909. In 1914, after the outbreak of the First World War, Dorothy (age four) and her younger sisters Joan (age two) and Elisabeth (age seven months) were taken back to England with their nurse, Katie Stevens. They stayed in a rented house in Worthing, close to John Crowfoot's parents, attended a local school and saw their mother only once over the next four years. At the end of the war, Molly, whose four brothers were all killed in the war, returned to England with a new baby, Diana, and took charge of the family. Katie had married and soon emigrated to Australia, which gave Dorothy increased responsibility as the oldest of the girls. Their father was briefly in England on leave, but when he returned to the Middle East (now working in Khartoum), Molly took the family to her haunts in East Anglia, where Dorothy and her siblings were home-educated until she was ten.

In 1920, they settled in the village of Geldeston, near Beccles in Norfolk, where Dorothy started to carry out chemistry experiments in one of the attics. She was particularly fascinated by the way crystals of things such as copper sulphate could be grown from solutions, and later wrote that she was 'captured for life by chemistry and by crystals'.† She used her pocket money to buy supplies

* During her career she was variously known as Dorothy Crowfoot, Dorothy Hodgkin and Dorothy Crowfoot Hodgkin. For simplicity, we shall stick to Dorothy.

† Quoted by Georgina Ferry from Hodgkin's unpublished memoir. Other unattributed quotes in this chapter are also from Ferry.

from the local chemist, where there were no restrictions on what might be sold to an independently minded ten-year-old child. The enthusiasm for chemistry was also expressed at school. Dorothy attended a state school in Beccles where, although the school was mixed sex, only boys were taught chemistry. This led to a confrontation, the outcome of which was that Dorothy and one other girl, Norah Pusey, were allowed to join the boys for chemistry, because they were both hoping to go to university and study science. Even in the early 1920s, this was still a rare ambition for a young girl, and it was even rarer that she was indulged in it. Her mother soon took to spending half the year in Geldeston and the other half in Khartoum, while the girls stayed with friends – a fairly normal pattern in the last days of the British Empire, at least for a certain stratum of society.

At the end of 1923, Dorothy and Joan went out to visit their parents, and stayed until the following Easter. Two things made a particularly deep impression on Dorothy during this trip. First, the chemist A.F. Joseph, who worked for the Wellcome Laboratory, was a friend of her father, and, impressed by her enthusiasm for chemistry, he gave her a portable surveying kit with all the equipment needed to carry out chemical analyses. Secondly, although the girls did not directly encounter the harsher aspects of British rule during their visit, their parents did not try to hide from them the truth about the violent way in which any protests were put down. Molly herself was deeply troubled by this, and was an early and active supporter of the League of Nations. In 1924, when the school held a mock 'general election', Dorothy was the only pupil willing to stand (unsuccessfully) as the Labour Party candidate. It was a sign of things to come.

Another pointer to the future came when Molly gave Dorothy the published version of the lectures for young people that William Bragg had given to the Royal Institution (part of the continuing tradition of Christmas Lectures at the RI). Bragg and his son Lawrence had pioneered the use of X-rays to study the structure of matter, and had shared the Nobel Prize for this work in 1915 – the only father and son combination to have been honoured in this way. Dorothy was fascinated by the idea that the patterns made by X-rays diffracted from the atoms and molecules in a substance could reveal how those atoms and molecules were arranged, for example in crystals, and especially the idea that this might provide insight into the structure of living things.

In order to fulfil her ambition of studying chemistry at Oxford University, she would have to achieve high marks in the School Leaving Certificate in March 1927. She worked intensively towards this in her last two years at school and achieved the highest overall mark of any female candidate, not only opening up the path ahead but providing her with a prize of £30. But she soon discovered that there was a snag. The next step would be an entrance examination for admission to Somerville College, and one of the examination subjects was Latin, which she had never studied. She also needed a second science to go alongside chemistry. The headmaster of the school she had just left gave her a crash course in Latin, her mother supervised her study of botany, and in March 1928 Dorothy passed both the written examination and an interview. She would become an undergraduate at Somerville in 1928, which gave her time to join her parents on an archaeological expedition in what is now Jordan. The chance came because John Crowfoot had become the director of the British School of Archaeology in Jerusalem in 1927, and had organised a dig on the site of some Byzantine churches.

Dorothy was an active participant in the dig, cleaning finds and making an accurate one-tenth scale drawing of the patterned mosaic pavements that were uncovered. But Oxford had not forgotten her. She received a letter from her tutor advising her that botany was not the best accompaniment to the chemistry course, and that she should have studied physics. When she casually replied that it might be hard to find physics books where she was, the tutor sent her a standard textbook, which she studied in her 'spare' time.

The young woman who began her studies in Oxford in October 1928 was far from an average eighteen-year-old: she had experience of research and a maturity which enabled her to take responsibility for her sisters in England while their parents stayed in the Middle East. Her parents also had important friends in Oxford, including Margery Fry, the principal of Somerville. And to ease her financial situation, she had an allowance of £200 a year. Dorothy was ready for Oxford – even if Oxford was still not entirely ready for women like her.

By 1928, women had been admitted to Oxford for half a century, and since 1920, unlike the situation in Cambridge, they had actually been awarded degrees. There were five women's colleges, which sounds impressive until you learn that there were 22 men's colleges. And although there were no restrictions on how many men could attend those colleges, in 1927 the University had passed a resolution limiting the number of women students, fixing it at 840 (the number they had that year) and setting a quota for each college. The fear of the male dons seems to have been that having too many women around would be a distraction for the male undergraduates. To that end, it was further stipulated that no new women's college would be allowed if it would take the number of female undergraduates above a quarter of the number of men.

Oxford was at that time a man's university with a quota of women, not a mixed university. Women had to have permission to entertain men to tea (which could not be taken in the woman's own rooms), could not visit men in their college rooms, needed permission to go boating on the river with men, and so on. The positive side to all this, as Georgina Ferry, Hodgkin's biographer, has pointed out, was that the women's colleges were passionate in their support of their students and had first-class tutors. With women only accepted in Oxford on sufferance, it was essential to maintain the highest academic standards to prove that they belonged – it was no use being as good as the men, they had to be better. And of all the women's colleges, Somerville had the highest academic standards. They only accepted the best students, and they expected them to work hard.

In the university as a whole, there were very few women studying science or maths. In the time that Dorothy was an undergraduate, in each year there were no more than 5 female students, with about 55 men. Chemistry at the university was still largely taught as an experimental science, what Ferry describes as 'cooking: follow the recipe and observe the result'. There was no theoretical explanation offered for why certain reactions occur and others do not, nor any attempt at understanding the three-dimensional structure of materials. But alongside her coursework Dorothy read up on her first love, crystallography, and the work using X-rays to probe the structure of matter going on at other places of learning. She also heard Niels Bohr give a talk describing his model of the atom when he visited Oxford. But the biggest impression on her came from another visitor, J.D. Bernal, from Cambridge University, who gave a talk on the way the properties of metals are determined by the way their atoms combine.

Dorothy worked long hours in the laboratory, not just on her coursework but on an analysis of glass tessera from mosaics found by her parents during their excavations. She also had to pay for her sisters' school fees and other expenses using the £200 sent by her father, look after them during the Christmas holidays, and generally act in loco parentis. Even her twenty-first birthday in May 1931 passed without any special celebrations; she was working hard in the lab in preparation for her final examinations. 'Final', however, is a misleading term in this context, since the exams were followed by a compulsory one-year course of research, called Part II. She decided to find a project in X-ray crystallography, supervised by Herbert Powell, always known as 'Tiny'. Her closest friend at Oxford, Betty Murray, noted that 'she is going to research into crystals with X-rays and is so thrilled wild horses wouldn't keep her from it'.

It was during her fourth year in Oxford that Dorothy would encounter another of the obstacles placed before female scientists. The chemists had an organisation called the Alembic Club, which arranged meetings with outside speakers that were free for all to attend. But they also had regular meetings for members only, to discuss the latest developments in the subject – and the club itself was exclusively for men. Even when Tiny Powell gave a talk about Dorothy's work at the club, she was not allowed to attend the meeting. But that snub still lay in the future, when she finished her exams.

There was time after finals for a visit to Germany, which was not yet, in 1931, in the grip of Nazism, and Dorothy was able to spend some time with the crystallography group in Göttingen before meeting up with her parents to visit Berlin. Back in Oxford in September, Dorothy shared a set of rooms with Betty Murray, who was appalled at the way Dorothy neglected herself and overworked, and took care to make sure that she did have proper meals. But she could not keep Dorothy out of the lab, where, Betty wrote, 'her one object is to stop up longer than her supervisor because she wants to work alone and is tired of being told exactly what to do and when'.

The entire Crowfoot family planned to meet up in Rome for Christmas in 1931, but even this one break in the working routine did not relieve Dorothy of responsibilities. She had to sort out passports for her sisters and take charge of the travel arrangements. Back in Oxford in the new year, Dorothy worked so hard that Betty became seriously concerned for her health. As the deadline for the submission of her Part II thesis loomed, on one occasion she worked almost continuously from 8.30am on a Wednesday until 11pm of the following day. Betty wrote 'naturally she looks like a ghost. She gets thinner and thinner … she has simply got to get the thesis in in the sixth week [of term] and she has so far written three pages'. When the thesis was finished, or so Dorothy thought, Tiny Powell said it was too short, and she had to rewrite it in more detail, then get it typed up, before submitting the final version, technically after the deadline, but she got it in to the relevant mail box before the box was opened on the following day. The result, of course, was a first-class degree. Dorothy was just the third woman to receive a first in chemistry at Oxford, and could now contemplate a career in research.

By now, Bernal was using X-rays to study biological molecules, and the prospect of working in this area fascinated her. To support her application to join him in Cambridge, one of her examiners wrote to Bernal that 'she thoroughly deserved her first … She has ambitions to work in the more purely organic compounds … She is quite a competent practical organic chemist … I hope you will be able to take her'. This was just one of several testimonials. In August 1932, Dorothy learned that Bernal had accepted her as a research student, and that she would be joining his team in October. Her aunt, Dorothy Hood, supported her financially, and she also had a £75 scholarship from Somerville.

Bernal was nine years older than Dorothy, having been born on 10 May 1901. He graduated from Cambridge in 1922 and worked with William Bragg at the Royal Institution in London before returning to Cambridge in 1927 as their first lecturer in structural crystallography. Outside of science, he was a committed communist, having joined the party in 1923, and he made his first visit to the USSR in 1931. At that time many British scientists and other

Dorothy Crowfoot Hodgkin

intellectuals were active in left-wing politics, partly in response to the rise of fascism and Nazism on the continent. Dorothy shared these views. Bernal's work involved analysing the patterns made by X-rays when they interact with crystals to work out the structure of the crystals – the X-rays actually interact with the electrons in atoms, and the regions where electrons are concentrated indicate where the atoms are. This was straightforward for simple crystals such as common salt (NaCl), but increasingly difficult as the number of atoms involved increased. Dorothy soon became his right-hand woman in this work, and at the urging of her colleagues registered to study for a PhD. In those days, the degree was not essential for a researcher, but curiously, although (as we have seen) Cambridge still did not award the BA degree to women in the 1930s, there was no barrier to female doctoral students. So Dorothy became a member of Newnham College as a registered PhD student.

Her work developed from Bernal's pioneering investigation of biological molecules. Although these molecules are not in the form of crystals in the body, in the lab they can be prepared as crystals then studied to reveal their structure. This is particularly important for biomolecules because the way they work in living creatures often depends on their shape. Bernal's speciality was the study of a group of compounds known as sterols, which includes, for example, cholesterol. So it was natural for Dorothy to choose sterols for her PhD work.

She had hardly settled into this work, however, when, early in 1933, a bombshell arrived from Oxford. Somerville wanted her back, as a research fellow. After some discussion, the final proposal was for a two-year fellowship, the first year of which she could spend in Cambridge, at £200 per year and with only limited teaching duties. It was also understood that this could lead to a permanent position – at a time when jobs in scientific research were extremely rare. But she would need equipment to work with. Through the Oxford professor of organic chemistry Robert Robinson, she obtained £600 from ICI to pay for the latest X-ray apparatus and everything else she would need, to be installed in the mineralogy department. The deal was sealed, and she could complete her two years in Cambridge with her future assured.

Bernal included Dorothy's name as co-author on many of the papers he published over the next few years, but she made only a minor contribution to the most significant of these. In 1934, Bernal obtained some crystalline samples of the protein pepsin,* contained in test tubes with some of the solution from which they had crystalised – the 'mother liquor'. He discovered that although the crystals lost their structure when they dried out, it was possible to take X-ray photographs while the crystal was still wet, sealed in its tube with the mother liquor. This was the first X-ray photograph of a crystalline protein, a spectacular achievement announced by Bernal in a joint paper with Dorothy in *Nature*. The reason why Dorothy did not play a bigger part in this work is because she had been suffering from severe rheumatic pain in her hands, and had been visiting a specialist in London the day Bernal obtained the crucial photograph. There was no treatment for this growing disability: all the doctor could prescribe was rest – but she was far too excited by the new discovery to take that advice.

In September 1934, Dorothy returned to Somerville, where as well as research she had to do some teaching (which she was very bad at, in contrast to her very clear writing style) and get involved in interviewing prospective students, which she hated because so many good candidates had to be turned away. But she now had her own lab, with equipment powered by electricity tapped off from the

* Proteins are key biomolecules, made up of sub-units called amino acids, which have been described as the building blocks of life.

physics department's 400-volt alternating current supply. As well as completing her PhD work on sterols, almost as soon as she was settled in Oxford she began studying crystals of insulin, with the long-term aim of determining the structure of this key biomolecule. It was indeed a long-term project – it only came to fruition at the end of the 1960s, when electronic computers became powerful enough to analyse the complex X-ray patterns from the crystal. But the first paper on insulin was published in *Nature* in April 1935 – the first to carry her name alone as author.

Dorothy's personal life at this time was complicated. Soon after leaving Cambridge, she seems to have begun a love affair with Bernal, who was a married man but never let that get in the way of his relationships with women. But in 1937 she met Thomas Hodgkin, a left-wing historian the same age as her, who had been working in Palestine for the British High Commission but became highly critical of the British suppression of the Arabs and support of political Zionism. He resigned and returned to England, where he joined the Communist Party. Bernal faded from the scene romantically, although he remained a good friend and scientific colleague, and Dorothy and Thomas were married on 16 December 1937. Marriage required her to formally resign her fellowship, which had only just been made permanent, but in line with convention it was immediately given back to her.

The year of her wedding had been significant in other ways. Dorothy had begun work on another protein, lactoglobulin, using the technique of photographing the crystals in their mother liquor; she had taken on her first Part II student, completed her own PhD, and she had been involved in a storm at Somerville when a female student who had been found in the rooms of a male student at Queen's College was expelled ('sent down'), while the young man was merely suspended for a term and allowed to complete his degree. Dorothy was the only fellow to protest at this unbalanced treatment of the 'miscreants' at a meeting of the college council. 'It's not that I think [it] a very meritorious case,' she wrote to a friend, 'I don't – but they're not bad.'

Her own married life hardly followed a conventional pattern. By this time, Thomas was working in Cumberland, lecturing on history for the Friends Voluntary Service for Unemployed Miners, and getting on well at the job. So she stayed at Somerville, and they got together, one way or another, most weekends. Alongside her work, Dorothy was active in the Oxford University Labour Party, where she served on the committee, and took part in debates and demonstrations about the growing political crisis in Europe. But these activities were limited by the fact that she became pregnant in the spring of 1938. Pregnancy hardly slowed her scientific work, however, and in November she gave a talk on her protein work at a meeting of the Royal Society. As she still used her maiden name on scientific papers at that time, and didn't bother with her doctorate, she appeared on the programme, eight months pregnant, as 'Miss Crowfoot', causing a few raised eyebrows.

She gave birth at her parents' home in Geldeston to a boy, Luke, on 20 December. Somerville offered to give Dorothy paid leave of absence for the first term of 1939, making her the first woman at the university to have paid maternity leave, as well as being the first fellow to give birth while holding that post, at a time when even marriage was rare for a female fellow. She needed every bit of that time, as she became ill after the birth and developed a severe attack of rheumatoid arthritis. She was still only 28. But she was able to find a temporary place to rent in Woodstock Road and was able to set up home there with the aid of a live-in nurse to look after the baby and a part-time cook. This stretched the family's financial resources to the limit, and Thomas was still based in Cumberland, but it meant that Dorothy could return to her research essentially full time as the acute attack of arthritis eased. When she did so, she found that her hands had been so damaged by the attack that she could not flip the main switch of the X-ray equipment; she had a piece of wood attached to the lever so that she could operate it.

In the summer, Thomas was in Oxford lecturing at a summer school in Balliol, and the couple had arranged a long-term lease on a flat in Bradmore Road, starting in September. Then the war broke out.

The main difference this made to their lives in the short term was that for safety, Luke was sent to stay with his grandmother, but he soon returned to Oxford. Dorothy's work was unaffected, and Thomas, who now had a lecturing job in Stoke-on-Trent with the Workers' Educational Association, was exempt from active service because of narcolepsy – an ailment which causes the victim to fall asleep without warning. The crystallography work actually benefited from the new situation, as Bernal's lab in London was closed and all its equipment, together with some of his team, transferred to Oxford. In order to pay for this brand-new apparatus, Bernal had applied to the Rockefeller Foundation for a grant, and after some bureaucratic wrangling they agreed instead to give Dorothy £1,000 for research 'on proteins and viruses', which was enough to pay for the equipment and fund the work for a couple of years. This was the beginning of a relationship between Dorothy and the Foundation which lasted into the 1960s and ensured that she could work without funding worries.

Under wartime conditions, the Foundation did not object when the focus of Dorothy's research was shifted to investigating the structure of penicillin, as part of a joint British–American project to develop techniques to produce the antibiotic in large quantities, which among other things transformed the survival prospects of battlefield casualties. Before that project got underway, on 23 September 1941 Dorothy gave birth to her second child, Prudence Elizabeth ('Lizzie'). This time, there were no complications and she was back at work in mid-October. It was not until

1945 that Dorothy was sure that she had determined the structure of penicillin, thanks to a breakthrough in the way data were analysed. The previous year, she had been able to get time on a Hollerith machine owned by the Medical Research Council; this was a precursor to modern computers that counted data from punched cards, greatly speeding up the analysis. By then, methods for obtaining penicillin from the penicillium mould were already operating on an industrial scale, but the Oxford work made it possible to develop an array of synthetic antibiotics based on the structure she had discovered. At the time, it was the largest and most complex molecule analysed using X-ray crystallography.

The significance of this breakthrough is highlighted by a famous story that Dorothy later recounted about an incident later in 1945:

> I remember sitting on the steps of the Royal Society ... talking to Bernal and I was telling him that we had solved the structure of penicillin. He said, 'you will get the Nobel Prize for this'. I said 'I would far rather be elected a Fellow of the Royal Society,' and he said 'that's more difficult'.*

He had a point, because the Royal Society had only elected its first two female fellows in 1944. But he was wrong. Dorothy was elected as a Fellow of the Royal Society in 1947, a year after the birth of her third child, Toby. This prompted a friend to write: 'If a woman FRS can have three children, anyone might do anything.'†

Throughout most of this period, Dorothy had no formal university appointment, and subsisted on her Somerville fellowship. In 1944 she was overlooked for the newly created post of reader in

* See H. Rose, *Love, Power and Knowledge*, Polity Press, Cambridge, 1994.

† See Ferry.

chemical crystallography (it went to Tiny Powell), and only managed to obtain the more junior post of demonstrator in May 1945. But if the prophet wasn't honoured in her own land, her American sponsors had a better idea of her worth. In 1947 the new FRS was awarded a three-month travelling fellowship by the Rockefeller Foundation, and arrived in New York on the *Queen Elizabeth* on 3 October, the first step on a tour of research laboratories that would take her as far as California, seeing 'all the X-ray single crystal photographs which have been taken of any protein crystals in the States'.

Thomas had held the fort in Oxford while Dorothy was away, but in 1948 it was her turn to look after things at home while he travelled in Africa advising authorities there on setting up adult education systems. This was the beginning of a career move which saw him increasingly involved with African history and African nationalism in the last days of the British Empire, with much more foreign travel, in which Dorothy was sometimes able to join as their children got older. But this work was initially not well paid, and the family struggled financially for years. Dorothy also struggled increasingly with mobility problems caused by her developing arthritis.

Against this background, in 1948 Dorothy began the work which would lead her to a Nobel Prize. By then, it had been known for a couple of decades that pernicious anaemia could be cured if patients were fed liver, and chemists had just succeeded in identifying the factor responsible or the cure, now known as vitamin B12, in the form of long red crystals. Chemists could find out which atoms were present in the molecules of the vitamin (carbon, hydrogen and so on), and in what proportions, but this information was not enough to explain how the atoms joined together to make specific shapes, which are important in biological activity. Dorothy was the natural person to ask to study the structure of these crystals.

By then, her laboratory was becoming increasingly well-equipped and well-funded, to the bafflement of the university authorities. When Bernal's equipment went back to London after the war, the Rockefeller Foundation funded replacements for her. She also received major grants from the Nuffield Foundation (£3,000 a year for five years to pay for computer time, some of it on the Pilot ACE computer developed by Alan Turing) and the Department of Scientific and Industrial Research. Instead of rejoicing in this recognition of her work, the university authorities worried that her lab might be getting 'unduly inflated' and suspiciously checked that this did not involve them in any 'concealed commitments'.

The vitamin B12 work pushed this equipment and the available computing power to the limit, and also involved some competition between different groups, detailed by Ferry. The final breakthrough came in 1954, when a colleague working at the University of California's Los Angeles campus was able to provide her with free computer time on what was then one of the most powerful computers in the world, and the discovery of the three-dimensional structure of vitamin B12 was announced in the summer of 1955.

Each molecule of the vitamin contains 181 atoms (its chemical formula is $C_{63}H_{88}CoN_{14}O_{14}P$), and the breakthrough pointed the way to the analysis of even more complicated biomolecules, including proteins. Lawrence Bragg, the doyen of X-ray crystallography, described the discovery as the equivalent of breaking the sound barrier, and Linus Pauling, who had received the Nobel Prize in 1954 'for his research into the nature of the chemical bond and its application to the elucidation of the structure of complex substances', wrote to 'congratulate you on the wonderful job you have done on vitamin B12. I find it hard to believe, although very satisfying, that the methods of X-ray crystallography can be used so effectively on such a complex molecule.' On the back of this recognition, Oxford

University appointed Dorothy as a reader in X-ray crystallography, with a salary of £1,650 a year, and in 1956 she became the first woman to be awarded the Royal Medal of the Royal Society – its highest honour. Her parents lived just long enough to enjoy her success – Molly Crowfoot died in 1957, and John the following year.

Almost as soon as she had completed her work on vitamin B12, Dorothy turned her attention back to insulin, a project that would occupy her through the 1960s. But where would she complete that work? By 1960, she had received several offers of professorships and was seriously considering leaving Oxford, where her status was still that of a mere reader. But then the Royal Society chose her as the beneficiary of a donation by Lord Wolfson, to fund a professorship, with no strings attached, providing a salary of £3,000 a year and a further £5,000 a year for research assistants and expenses. She became the first Royal Society Research Professor in the University of Oxford, relinquishing her readership in the process, and causing some more alarm among the university bureaucracy. They rather grudgingly gave her the honorary title of university professor, but since she was no longer on the university payroll they pointed out that she no longer had the right of free access to the computer service. After some discussion, they decided to allow her up to a thousand hours a year. But, bureaucracy notwithstanding, she was now essentially free to work as she pleased and when she pleased.

In 1964, Dorothy was awarded the Nobel Prize for Chemistry, the first (and so far only) British woman to receive the honour. Indeed, she was only the third woman to receive the chemistry prize, joining Marie Curie and her daughter Irène Joliot-Curie. The citation read 'for her determinations by X-ray techniques of the structures of important biochemical substances', but the press made much of the fact that Dorothy was a 'mother of three', and one report described her as a 'housewife', which seemed to impress them more than the fact that she was a Fellow of the Royal Society. It was perhaps this kind of coverage which led her to say in a speech at a banquet following the prize ceremony: 'The situation in which I find myself will, I very much hope, not be so uncommon in the future that it will require any comment or special treatment – as more use is made of the many gifts which women share equally with men.'

Most of the prize money was used over the years to provide help for graduate students at Somerville, including start-up funds for a college nursery. Other awards and honours inevitably followed the Nobel Prize, the pinnacle being the Order of Merit, Britain's highest honour, which is the personal gift of the Monarch. There are only ever 24 members of the Order, so someone literally has to die before a new member is appointed. In 1965, there were two vacancies, caused by the deaths of Winston Churchill and T.S. Eliot; they were filled by Dorothy and Benjamin Britten.

While all this was going on, Dorothy was in charge of a large research group, whose main focus was on the insulin problem. Although she was inevitably less hands-on than in the vitamin B12 work, it was very much her project, and it was at the back of her mind that in 1970 she would reach the usual retirement age at Oxford. With the aid of improved X-ray cameras and more powerful computers, in July 1969 the team achieved their goal of mapping the position of every atom in the molecule. It contains 788 atoms, forming two chains of amino acids which are linked together by sulphur atoms. The formal announcement was made at the Eighth International Congress of Crystallography, held a month later in New York. It was typical of Dorothy that she chose the most junior member of the team, Tom Blundell, to give the lecture reporting the breakthrough. It was also typical that she did not regard this as the last word on the subject, but continued to work on refining the accuracy of the map long into 'retirement', with her final paper on insulin appearing in 1998, when she was 78 years old. The determination of the structure of insulin pointed the way for it to be manufactured and used on a large scale for the treatment of diabetes.

Her 'retirement' also involved travel in China, Africa and India, promoting science education and world peace, initially often with Thomas, who died in 1982.* The last of her scientific trips was to the 1993 meeting of the International Union of Crystallography, held in Beijing, which she attended against medical advice (she had broken her hip in a fall in 1991 and was both frail and confined to a wheelchair), but enjoyed the last chance to mingle with her scientific friends. The following year, after another fall, she never regained consciousness and died at home on 29 July 1994. At her memorial service, Max Perutz, another Nobel Prize-winning crystallographer, said: 'She radiated love: for chemistry, her family, her friends, her students, her crystals and her college … Her love was combined with a brilliant mind and an iron will to succeed, regardless of her frail and later severely crippled body. There was magic about her person.'

In 2010, to celebrate the 350th anniversary of the founding of the Royal Society, Dorothy was the only woman featured in a set of stamps commemorating ten of the Society's most illustrious members.

* She was awarded the Lenin Peace Prize by the Soviet government in 1987 in recognition of this work.

CHAPTER 9

Chien-Shiung Wu (31 May 1912–16 February 1997)

Discovered the law which allows matter to exist in the Universe today

The Chinese-born American physicist Chien-Shiung Wu is the odd one out in our gallery of female scientists. Although she did have some relatively minor hurdles to overcome in order to pursue a career as a physicist in a male-dominated discipline, these were not as great as the problems faced by many of the other women featured in this book. And yet, when it came to the biggest hurdle of them all, Wu was denied the accolade she deserved, almost certainly because of her gender.

The Nobel committees have made many controversial – and even bizarre – decisions over the years, but one of their biggest blunders, certainly in science, has been their failure to acknowledge the work of Wu. For this reason alone, she deserves her place in these pages. You may feel, rightly, that Lise Meitner was hard done by. But in her case the confused circumstances surrounding her flight from Germany and the omission of her name from key publications does provide some sort of explanation, if not really an excuse, for the omission. However, in Wu's case there is no confusion, no explanation and certainly no excuse. Her experiments were certainly worthy of the prize – they provided an explanation for why the Universe as we know it exists

This chapter, then, will concentrate on the science, with less emphasis on the biography, which is appropriate for another reason. In Wu's life, work came first and family a distant second. In an essay about her grandmother, Wu's granddaughter Jada Yuan confessed: 'I have to confront the idea that, in the course of her many achievements, Chien-Shiung Wu didn't balance her work and her family life, and those choices have trickled down, through my father and then to me.'* One biographer† quoted her as saying that 'a nice husband, a short commute and good child care [were] prerequisites to being a successful woman in science'. A pattern that fits, to a greater or lesser degree, many of our subjects.

Wu was born in Liuhe, near Shanghai, in the province of Jiangsu, on 31 May 1912. Her parents had radical ideas for the time, reflected in their choice of her names. The family tradition was that all boys born in her generation would have the name linked to the characters for 'Chien', followed by characters from a phrase 'Ying-Shiung-Hao-Jie', which can be translated as 'outstanding hero'. So their son – Wu's older brother – was named Chien-Ying Wu, and when their daughter was born, in the interests of gender equality, the parents went against tradition and gave her the next 'male' name, Chien-Shiung. A later brother duly became Chien-Hao. With so many Chiens, we shall stick to calling 'our' Chien by the name Wu.

* https://www.washingtonpost.com/lifestyle/2021/12/13/chien-shiung-wu-biography-physics-grandmother/

† Sharon Bertsch McGrayne, *Nobel Prize Women in Science: Their Lives, Struggles, and Momentous Discoveries*, Joseph Henry Press, Revised edition, 1998.

Wu's father, Wu Zhong-Yi, was an engineer and teacher who had been actively involved in Sun Yat-Sen's revolution, which overthrew the Qing dynasty and led to China becoming a republic in the year Wu was born. Her mother, Fan Fu-Hua, was a teacher and an advocate for gender equality. After the revolution, Wu Zhong-Yi gave up his work as an engineer to found a school in Liuhe which provided education for girls from all social backgrounds, including his daughter. At the age of ten, she moved on to a boarding school in Suzhou, 50 miles away, and in 1929 entered the competitive national examination for teacher training college, where she came ninth out of some ten thousand applicants. This high placing meant that after a year spent teaching, she could enrol at what was then the National University (now Nanjing University), one of the oldest and most prestigious institutions of higher learning in China, to study mathematics and physics; she graduated in 1934.

Like her father, Wu was active in politics at this time, leading student protests about what was seen as China's conciliatory attitude to Japanese aggression. But she did not neglect her studies, and after graduating was able to spend two years in further study and research supervised by Gu Jing-Wei, a female physicist who had a PhD from the University of Michigan – at that time, it was not possible to work for a PhD in fundamental physics in China. This first experience of experimental research was in X-ray crystallography, but Wu's career then took a different tack. Following in Gu Jung-Wei's footsteps, she applied to Michigan and was accepted as a PhD student, with an uncle, Wu Zhou-Zhi, sponsoring her financially. She sailed for the United States on board the steamship *President Hoover* in August 1936, when she was 24. Wu never saw her parents again, and it would be almost 40 years before she saw China again.

Almost as soon as Wu arrived in America, her plans changed. The ship docked at San Francisco, and she took the opportunity to visit the physicists at the nearby campus of the University of California, Berkeley. The director of the radiation laboratory there was Ernest O. Lawrence, who had invented a type of particle accelerator called a cyclotron, for which he would be awarded the Nobel Prize for Physics in 1939. Such accelerators are crucial to the investigation of sub-atomic particles, because by colliding charged particles such as protons and electrons at close to the speed of light they make showers of new particles, literally created out of the energy of the collision in line with $E = mc^2$; in this way, they manufacture the short-lived particles which physicists study. Wu was shown around the radiation lab by a 24-year-old PhD student, Luke Chia-Liu Yuan,* who had recently moved to America and who seems to have made as much of an impression on her as the cyclotron. Thanks to the cyclotron, Berkeley was a world leader in nuclear physics, which made Michigan seem less attractive. When Wu learned that the University of Michigan was so hide-bound by sexism that women were not even allowed to use the front entrance of the student union, it seemed even less desirable, and she persuaded the physics department at Berkeley to take her on as a research student. Yuan transferred to the California Institute of Technology, which offered him a scholarship, but this did not stop the development of his relationship with Wu.

Lawrence was Wu's official supervisor at Berkeley, but in practice she worked with an Italian émigré, Emilio Segrè, on aspects of beta decay, the process in which an atomic nucleus spits out an electron and is transmuted into a nucleus of a different element. They were able to identify all the elements in the chain of radioactive decay following nuclear fission, including in particular the behaviour of the element xenon, which would soon turn out to

* Yuan was a grandson of the second president of the Chinese Republic, Yuan Shikai.

have implications for the Manhattan Project – the development of the first atomic bomb.

Wu's ability was widely recognised at this time, but the praise of her contemporaries, when looking back now on those years, often carried an unconscious flavour of sexism. To Lawrence, she was 'the most talented female experimental physicist he [had] ever known', while the future Nobel laureate Luis Alvarez, who had the office next door to hers at Berkeley, described her as 'the most talented and most beautiful experimental physicist I have ever met'. Had she been a man, would they have commented on her gender and beauty? Jada Yuan said: 'Almost anything written about her at that time refers to how pretty she was, Orientalist in a leering way, as if shocked that she could also be the person J. Robert Oppenheimer deemed 'the authority' on beta decay.' She was also often compared favourably, in terms of her scientific ability, with Marie Curie. But why not with, say, William Bragg or Ernest Rutherford?

The first part of Wu's PhD thesis, completed in 1940, dealt with the nature of the radiation produced when beta-decay electrons were accelerated by passing near nuclei of xenon (bremsstrahlung radiation). The second part, building on the new discoveries made by Lise Meitner and her colleagues in Germany, dealt with the way radioactive forms of xenon, potentially useful in medical applications, could be produced by splitting uranium nuclei. The examiners for her PhD included Ernest O. Lawrence and Robert Oppenheimer, who found that she was an expert in understanding the way neutrons are absorbed by nuclei; something he would not forget when he needed such an expert later. Wu had not intended to stay in the United States – she planned to work there for a short time before returning to China – but the Japanese invasion of China in 1939 encouraged her to delay her return. At that time, none of the top twenty research universities in the United States had any women on their physics faculty, so Wu was not too surprised or concerned when she failed to obtain a long-term post at a university. She spent the next two years as a post-doctoral fellow at the Radiation Lab. The outbreak of war between the USA and Japan in December 1941 made a return to China impossible, and she ended up travelling in the opposite direction, to the eastern United States, and eventually breaking through some of the barriers to female physicists.

Luke Yuan had also completed his PhD, and Wu married him on 30 May 1942, before the couple moved to Massachusetts, where Yuan had secured a job working on radar at RCA (Radio Corporation of America). This was the first and only time she conformed to the female stereotype by following in the wake of her husband,* and initially it was not a satisfactory move for her in career terms. The only job Wu could get was as a teacher in a women's college (Smith College), where there was no opportunity for research. It was only after intense lobbying on her behalf by Lawrence that she became an 'instructor' at Princeton University – the first female faculty member in the department of physics. This was an even more impressive breakthrough than it looks at first sight. Princeton then had an all-male student body – the first female students were only accepted in 1969.

In March 1944, Wu was recruited to the Manhattan Project, initially working with a team at Columbia University in New York City on developing a way to enrich weapons-grade uranium with the radioactive component of the element, by separating fissionable U-235 from stable U-238. Jada Yuan tells how the recruitment process went:

* She remained known as Dr Wu, and was quick to correct anyone who referred to her as 'Mrs Yuan'.

> Two physicists in Columbia's division of war research spent a day quizzing her, but avoided telling her anything about the work she'd be doing. Then they asked her to guess. 'I'm sorry, but if you wanted me not to know what you're doing, you should have cleaned the blackboards,' she said.

They hired her on the spot.

Later that year, the scientists working with the first practical nuclear reactor, at a location in Washington State known as the Hanford Site, found that it was unexpectedly shutting down. They suspected that the isotope xenon-135, produced by the fission of uranium, might be causing the problem (dubbed 'xenon poisoning'), and recruited Wu, the world expert on radioactive xenon, to look at the problem. Her calculations, jumping off from the experiments carried out for her PhD, showed that the xenon was very efficient at absorbing neutrons (it has a large 'neutron absorption cross-section'), trapping neutrons before they could build up a chain reaction in which neutrons released by the splitting of one nucleus trigger the fission of more nuclei. Along the way, she designed an improved version of the Geiger counter, the instrument used to measure radiation. Armed with Wu's data, the physicists were able to calculate how much more plutonium they would need to produce a sustained chain reaction and make the reactor work. This work was a key part of the programme to develop enriched uranium for atomic bombs; like many of her colleagues, in later years Wu regretted this and campaigned against the spread of nuclear weapons.

When the Second World War was over, the civil war in China and the victory of Mao Zedong's communist faction meant that it was still impossible for Wu and Yuan to visit China. At that time, the United States authorities would not issue a new entry visa to anyone who had been to a communist country, so if she did go back to China it would have to be permanently. If she did ever consider doing this, her mind was made up when she received a letter from her family in which her father urged her not to go back.* Instead, in 1945, she moved permanently to Columbia University, initially as a research professor, a non-tenured post, but in 1952 she would be appointed as a tenured associate professor, their first female tenured professor of physics.

Her family life settled into a stable pattern between those years. Her son Vincent was born in 1947 (one of the visitors who came to the hospital to congratulate Wu and meet the new baby was Albert Einstein), and Yuan got a post at the Brookhaven National Laboratory on Long Island, New York. Wu stayed at Columbia for the rest of her career, where the first significant research she carried out after the birth of her son confirmed the accuracy of the theory of beta decay, building on her earlier work at Berkeley. This established her as the leading expert on the subject. She was at Columbia when she carried out the work which should have led to the award of a Nobel Prize. But immediately after the beta-decay work, she performed an experiment which now seems of almost equal importance, although it was largely unsung at the time.

The theory of quantum mechanics, or quantum physics, describes the behaviour of things on very small scales, including the behaviour of entities such as electrons and photons (the 'particles' of light). The theory had only been completed in the late 1920s, following the breakthrough work of Erwin Schrödinger and Werner Heisenberg, and in 1935 Albert Einstein had drawn attention to a bizarre prediction of the theory.

* In the late 1940s, Wu and her husband were actually invited to return permanently to China to work at the National Central University, but they heeded her father's advice.

Chien-Shiung Wu
Smithsonian Institution, via Wikimedia Commons

Under certain circumstances, a quantum system, such as an atomic nucleus, can emit two particles in opposite directions. The quantum rules say that the properties of these particles are correlated. For example, there is a property called spin. If one of the particles has spin up, the other must have spin down. So far, so good. But Einstein emphasised that the rules also say that the property in question does not become real until the particle interacts with something else. At that moment, particle A 'decides', in some sense, that it has, say, spin up. At the same instant, particle B must therefore settle into a state with spin down, even though it may be far away from particle A. Einstein, who called this 'spooky action at a distance', said that it was impossible, so the 'prediction' proved that quantum theory was flawed. But Schrödinger, responding to Einstein, gave it the name 'entanglement' and said that it was real. Today, entanglement is seen as a fundamental feature of the quantum world – perhaps the defining feature of quantum physics – and it has practical applications in, for example, quantum computing. It only became accepted in this way in the 1980s, after several experimenters proved conclusively that it is real. Only later did physicists realise that Wu had already provided persuasive evidence for the reality of entanglement in 1949.

Wu, working with her research student Irving Shaknov, had decided to test a prediction made by the theorist John Wheeler. This involved using a cyclotron at Columbia University to create collisions between electrons and positrons. When these particles meet, they are annihilated in a burst of energy, with photons flying off in different directions. These photons each have a property called polarity, which can be represented as a pointer like the hour hand of an analogue clock, pointing in any direction around a circle. Wheeler's calculations said that pairs of these photons should have correlated polarity – at right angles to each other, with one pointing 'up' (12 o'clock) and the other 'across' (3 o'clock). The experiment showed exactly this correlation, not only confirming Wheeler's prediction but also, because of the way the experiment was set up, being consistent with the presence of entanglement, although it was not definitive proof that entanglement was at work.

But when they published their results in the *Physical Review* in January 1950, Wu and Shaknov did not mention entanglement, only the confirmation of Wheeler's prediction. It may be that Wu didn't consider entanglement worth mentioning, or that she was reluctant to get involved in what was then still a controversial topic. But it was left for a later generation to finally prove the reality of quantum entanglement, with definitive experiments designed specifically to test for its effects. In 2022, three physicists – Anton Zeilinger, Alain Aspect and John Clauser – shared the Nobel Prize for this work. The work that should have earned Wu her own Nobel Prize was carried out in the mid-1950s, after some more changes in her personal life.

After the communists came to power in China in 1949, Wu's passport, issued by the now defunct Kuomintang government, had no validity. Her ability to travel outside the United States was severely limited, although she was granted special permission to visit colleagues in Switzerland. Politically, she offered lukewarm support to the rump of the former Republic of China, confined to the island of Taiwan but claiming to be the rightful rulers of the whole country. This evidence that at the very least she did not support the communist regime probably helped her successful application to obtain US citizenship in 1954, when anti-communist feeling was running high. So it was as an American citizen that Wu carried out her most important scientific work.

One of Wu's colleagues at Columbia was the theoretical physicist Tsung-Dao Lee, who had also been born in China, in 1926, but moved to America in 1946 and would become an American citizen in 1962. He had been working with Chen-Ning Yang, who had been born in China in 1922, also moved to America in 1946, and in the 1950s was working at the Institute for Advanced Study, in Princeton. Yang would become an American citizen in 1964. Lee and Yang had been puzzling over the strange behaviour of two kinds of particle, at that time known as theta and tau, part of a family called mesons. The properties of the theta and tau particles (things like electric charge, and mass) were essentially identical, and they could only be distinguished because of the way they decayed into other particles. The obvious explanation was that there was only one kind of particle, which could decay in two different ways. But the specific details of these particular decays meant that in order for this to be possible a fundamental property of nature called parity conservation had to be violated.

The simplest way to understand this is in terms of reflection. Parity conservation means that if you carry out an experiment and imagine carrying out the same experiment in a mirror-image Alice-in-Wonderland world, the two experiments will give the same results. Parity non-conservation is when the mirror experiment gives a different result from the one in the lab. Of course, you can test this by running a mirror-image version of the original experiment. By the mid-1950s, many experiments had shown that this property is conserved in experiments that involve electromagnetism and in those that involve the so-called 'strong' force, which holds atomic nuclei together. But nobody had actually tested the assumption that it also holds for processes involving the 'weak' interaction, which among other things governs the behaviour of particles during beta decay and regulates the forces that allow the Sun to keep shining.

Following a scientific meeting in April 1956, where the puzzle of the theta and tau particles was discussed, Lee and Yang realised that the particular decay modes of theta and tau could be explained if they were identical particles but could decay in ways which violate parity conservation. Their theoretical colleagues regarded this as an outlandish idea, hardly worth taking seriously. But Lee knew Wu, who at 44 was established as the world's expert on beta decay experiments. So in 1956 he asked her if she could devise the necessary experiments to test the theory he had developed with Yang. She could.

The experiments involved studying the beta decay of radioactive cobalt-60 atoms cooled to very low temperatures (close to absolute zero, -273 degrees Celsius) and held in a uniform magnetic field. The experiments were actually carried out at the National Bureau of Standards (now the National Institute of Standards and Technology) in Washington, DC, which had the technology for work at temperatures close to absolute zero. Wu commuted between New York City and Washington for several months to oversee the day-to-day work of her junior colleagues. The results showed that the way electrons (beta particles) were emitted violated the 'law' of conservation of parity, and confirmed the predictions made by Lee and Yang.* There were not two separate kinds of particle, the theta and tau, but a single type, dubbed the kaon, which could decay in ways which violated parity conservation – parity non-conservation applied in some circumstances for the weak interaction. This was a scientific bombshell. It meant that nature is not perfectly symmetrical. This was a key step towards developing the 'standard model' of particle physics, but it is also a fundamental part of the standard model of the origin of the Universe.

* Specifically, if parity was conserved, the beta particles (electrons) expelled by the decay of cobalt-60 should be emitted more or less equally in all directions. What Wu found was that more particles flew off in one direction, opposite to the direction of spin of the nucleus. So parity was not conserved during beta decay.

The matter that makes up the everyday world, everything we can see, feel or touch, is made of atoms, which are themselves composed of particles such as protons and neutrons. Experiments with high-energy particle accelerators show that each kind of particle has a counterpart 'antiparticle'. A positron, which is a positively charged counterpart to the negatively charged electron, is an example, and the positively charged proton has a counterpart with negative charge, the antiproton. But there are no free antiparticles in the world at large. In line with Einstein's famous equation, a particle–antiparticle pair (such as an electron and a positron) can be made out of pure energy in cyclotrons and their successor accelerators. But when a particle meets its antiparticle counterpart (for example, an electron meeting a positron) the pair annihilate in a burst of energy.

The standard model of the Universe, based on a wealth of reliable observations, tells us that about 13.8 billion years ago the Universe was in a hot, dense state, a fireball of energy called the Big Bang (how it got to be in that state is still a matter of debate). In that state, energy was constantly being converted into particle–antiparticle pairs, and particle–antiparticle pairs were constantly being converted back into energy. As the Universe cooled, the process of pair creation stopped, and the leftover particles and antiparticles remained. If there had been an exactly equal number of particles and antiparticles, over time they would all have annihilated each other, with no matter left over. But because of the tiny asymmetry in the Universe related to parity non-conservation, for every billon antiparticles made in the Big Bang there were about a billion and one particles. When all the annihilation had finished, the billionth of the matter in the form of those excess particles was left over to make stars, planets and people. And it was Wu who first measured this effect. Her experiment explains why there is something rather than nothing, and ranks in importance with the

Michelson–Morley experiment which showed that the speed of light is a constant, always the same no matter where it is measured from or how the person doing the measuring is moving.

Wu's results were published in February 1957,* and the importance of the work is highlighted by the fact that the very same year it was recognised by the award of the Nobel Prize in Physics – to Lee and Yang alone. Wu was not included, even though the rules allow three people to share a Nobel Prize, and even though Alfred Nobel's will specifically says that the prize should be awarded 'to the person who shall have made the most important discovery or invention within the field of physics', which clearly requires experimental proof. Without the experimental proof, Lee and Yang would merely have been the authors of a speculative hypothesis, and they would not have got the prize. The only plausible explanation for Wu's omission is that she was a woman.† Clearly, in this case there can have been no question of racial bias, and her career up to that point had been more distinguished than those of either Lee or Yang. They both acknowledged her contribution in their Nobel speeches, and were among the many people who nominated her for the Prize in the following years – until in 1966 the Nobel Committee decided to stop publishing the list of nominees 'to avoid controversy'. But the mistake – the biggest one in the history of the Nobel Prize – was never rectified.

At Columbia, Wu's achievement was quickly recognised by her appointment as a full professor in 1958; the same year she was elected as a member of the US National Academy of Sciences

* C.S. Wu, E. Ambler, R.W. Hayward, D.D. Hoppes and R.P. Hudson 'Experimental test of parity conservation in beta decay' *Physical Review*, volume 105, page 1413.

† The irony that her work concerned the breakdown of parity has often been pointed out.

(the seventh woman to receive that honour) and became the first woman to receive an honorary doctorate from Princeton University, where the citation called her the 'top woman experimental physicist in the world'. But in spite of all this, she was not paid on the same scale as her male colleagues until 1975! Over the next few years, Wu and her colleagues carried out experiments to test an improved theory of beta decay, incorporating an idea from Richard Feynman and Murray Gell-Mann called conserved vector current, or CVC. This proved a crucial step along the road that led to a theory called the electroweak interaction, which combined electromagnetism and the weak interaction in one mathematical package. This in turn became a key component of the so-called 'standard model' of particle physics, which also incorporates the strong force. She then wrote a textbook titled *Beta Decay* with her colleague Steven Moszkowski, published in 1966; it was the definitive book on the subject for the rest of the century, and is still used today.

But Wu's growing reputation did not open the way for a return to China. Her older brother had died in 1958, her father in 1959 and her mother in 1962, but as a US citizen she had been unable to attend any of their funerals. The closest she got was a visit to Hong Kong in 1965, where she met up with her younger brother and the uncle who had originally sponsored her move to America. It was only after the relations between China and the USA thawed in the 1970s that she was able to make several visits to both mainland China and Taiwan, but by then her uncle and brother had been killed in the Chinese Cultural Revolution. Wu herself had become an outspoken advocate for civil and women's rights. In 1964, at a symposium on women in science and engineering held at the Massachusetts Institute of Technology, she delighted her audience with the comment: 'I wonder whether the tiny atoms and nuclei, or the mathematical symbols, or the DNA molecules have any preference for either masculine or feminine treatment.'

She continued: 'In our present society of plenty and proficiency, is it too much to provide excellent professional child care during the day so that mothers can get away from monotonous household chores and work in their chosen fields?' But while acknowledging that it is important for scientists to have home lives, she concluded: 'However, this noble human desire to be devoted companions and parents must, ideally, be equally shared by men.'

In 1975, Wu became the first woman to be president of the American Physical Society, and she used the position to persuade President Gerald Ford to create the Office of Science and Technology Policy. The same year she was the first woman to receive the National Medal of Science in Physics; the National Medal is the highest honour for American scientists. And in 1978 she was the first recipient of the Wolf Prize in physics, which was created partly to recognise people who deserved the Nobel Prize but hadn't actually received it.* The citation referred to 'her explorations of the weak interaction, helping establish the precise form and the non-conservation of parity for this natural force'. Altogether, she received sixteen honorary degrees, had an asteroid named after her (the first living scientist to receive this honour) and appeared on an American stamp issued on 11 February 2021, to mark the International Day of Women and Girls in Science. Wu retired in 1981, at the age of 69, but remained an honorary member of the faculty at Columbia until she died on 16 February 1997. In 'retirement' she promoted the scientific education of girls in the USA and China, and offered herself as a role model by giving talks about her life and career. Her ashes were buried in the courtyard of the school, founded by her father, that she had attended as a girl. Her husband died six years later, on a trip to China.

* Since 1978, many of the recipients of the Wolf Prize have *later* been awarded the Nobel equivalent.

Perhaps Wu's best epitaph is the citation she received in 1982 when awarded the Science for Peace prize from the Italian Ettore Majorana Centre for Scientific Culture: 'for her intense and vast scientific activity that has permitted the understanding of weak forces and for her engagement in the promotion of the role of women in science.'

CHAPTER 10

Rosalind Franklin (25 July 1920–16 April 1958)

Used X-rays to determine the structure of viruses

Rosalind Franklin (known in conversation to her family as 'Ros', and to others as 'Roslind', without pronouncing the 'a') had it comparatively easy early on, coming from an affluent background and with nobody objecting to her choice of career. It was only when she had reached a certain level of success in that career that she encountered obstacles because of her gender, and had to fight for the same level of recognition and reward that came automatically to her male colleagues. Her early death, at the age of 37, left a confusing legacy in which she was first ignored, then cast in the role of a bit player in one of the most important scientific findings of the twentieth century, then hailed as a heroine and feminist martyr as the pendulum swung the other way. The truth, of course, is more nuanced, and among other things shows that she was no scientific one-trick pony, but made major contributions both before and especially after her role in the discovery of the structure of DNA. The irony is that although she is now most usually remembered for that DNA work, what she did next was actually her greatest achievement, and itself only the precursor to what might have been. Her story would be worth telling even without that interlude in her career.

Franklin came from a prominent Anglo-Jewish family. They were involved in merchant banking and the publishing house of Routledge & Kegan Paul, and her great-uncle Herbert Samuel was a politician whose career has (unfortunately) influenced events in the Middle East to the present day. As home secretary in 1916, he wrote a memorandum which formed the basis of the Balfour Declaration of 1917, in which the foreign secretary Arthur Balfour proposed 'the establishment in Palestine of a national home for the Jewish people'. When Britain emerged as the de facto rulers of Palestine after the First World War, Samuel became the first high commissioner of the territory and was able to begin putting theory into practice.

Ellis Franklin, Rosalind's father, had served in the war, married Muriel Waley in 1917, and joined the family business of Keyser's bank the following year. He was the youngest of six children, and with Muriel produced five of his own, starting with David in 1919. Rosalind Elsie, born on 25 July 1920, was the second child, followed by Colin in 1923, Roland in 1926 and Jenifer in 1929. The family lived in comfort in a large house just off Westbourne Grove in London. There was a nanny, Ada Griffiths, who joined the family when David arrived and stayed until after the children were all grown; she was not regarded as a servant, but there were plenty of those to ensure the smooth running of the establishment. Franklin's biographer, Brenda Maddox, describes the family as 'frugal rich', their one indulgence being foreign travel. Charitable work was also important, and Ellis Franklin often lectured on scientific topics at the Working Men's College near King's Cross station.

Rosalind Franklin's own education was very much in line with this lifestyle. When she was six, she started at a private day school within walking distance of home, but at the age of nine moved to a boarding school at Bexhill, on the Sussex coast. Ostensibly, this was because she had suffered from minor childhood ailments and it was thought that the sea air would be good for her health;

however, as her sister Jenifer was born on 5 November 1929, her parents may have been happy to have one less child for the nanny to worry about. She remained at the school for two years, becoming self-sufficient and doing well both academically and at sports. Then, in January 1932, she moved on to St Paul's girls' school, in Hammersmith, a bus ride away from the family home. This was a counterpart to the long-established St Paul's school for boys, but there was no interaction between the two establishments.

St Paul's was a popular choice for Jewish families because it had no religious connections, and the Franklin family were sufficiently relaxed about tradition that they did not object to Rosalind taking part in sports on Saturdays, the Jewish sabbath. She was in the school teams for tennis, hockey and cricket, while also excelling at academic work, particularly science and languages (German and French). After four years, when she joined the sixth form to study for the Higher School Certificate she chose to specialise in physics, chemistry and mathematics. But her sixth-form studies would be completed against the background of the developing political situation in Europe, which was particularly concerning for Jewish families.

Both of Rosalind's parents were actively involved in settling Jewish refugees in Britain, with Ellis Franklin even cutting down his time at the bank to work with a refugee committee. In 1938, the Franklins took two German children in to live with them. Rosalind was also involved, as each day some of the girls at St Paul's were sent to help with office tasks for the committee Ellis worked with. None of this affected her school work, where she regularly won prizes and was usually top of her class. In February 1938, still only seventeen, she took the entrance examination in physics and chemistry at Cambridge, and was interviewed by Newnham and Girton colleges, each of which offered her a place. A few months later, she achieved six distinctions in her Highers and accepted the offer from Newnham. She was also awarded a scholarship worth £30 a year, but at the suggestion of her father passed this on to an impoverished refugee student. The award was because she had come first out of all the applicants in the Cambridge entrance examination in chemistry.

With her future assured, Franklin did more work for the refugees flooding in from Austria following its annexation by Germany, and found time for a holiday in Paris. But not long after Neville Chamberlain flew back from his meeting with Hitler to proclaim 'peace in our time', and only a few weeks after her eighteenth birthday, she was off to Cambridge, where life for female students was still very much as we have seen in earlier chapters. Franklin took her work seriously, signing up to study physics, chemistry, mathematics and mineralogy, plus scientific German. As a member of a society called the Archimedeans, she went to lectures by J.J. Thomson and J.B.S. Haldane, and also attended a meeting of the Association of Scientific Workers where William Bragg was the chair. Sports featured largely in her life – particularly squash and tennis – and she cycled everywhere, including, to the astonishment of her parents, home to London at the end of term. This left little time for socialising, which she was not inclined towards anyway, and she made few close friends at college.

It's a sign of the times that in January 1939 in a letter home Franklin described one of her lecturers as 'very good, though female',* but just four months later, on 6 May, the archaeologist Dorothy Garrod was appointed as the first female professor in the history of Cambridge University. This caused consternation among the colleges. Part of their tradition was that a newly appointed professor would be invited to a grand dinner – a 'feast' – at each of the colleges in turn. But the men's colleges

* Quoted by Maddox.

had a rule that no women were allowed at feasts. King's College, to its credit, broke the rule and invited Professor Garrod; the rest, to their shame, did not.*

Cambridge was still insulated from the real world in other ways. Franklin was shocked by the pacifist inclinations of many of her fellow students, and by the reluctance of the government to admit more refugees from the Nazis. But in May 1939, after a short holiday hiking in the Peak District, she had to buckle down to her preliminary exams, in which, still a couple of months short of her nineteenth birthday, she got a first. That summer the entire family went on holiday to Norway, where news of the Soviet–German non-aggression pact of 24 August 1939 brought an end to the vacation and a hurried return home. It was obvious to Ellis Franklin what would happen next, and Germany duly invaded Poland on 1 September.

Rosalind's brother David had just completed two years at Oxford, but he gave up his course to join the army. Rosalind, after some discussion with her father about whether she should also find work relevant to the war, returned to Cambridge, where among other things she read Linus Pauling's book *The Nature of the Chemical Bond*, the first to explain how atoms are held together to form molecules based on the principles of quantum physics. And she began to learn about X-ray crystallography. At the end of the year, she got a first in her 'Part One' exams, permitting her to move on to 'Part Two' immediately in her third year, and earning a scholarship worth £15.

* When we were at Cambridge in the 1960s, some colleges had a rule that on such occasions one could invite a female guest 'if she were the sort of woman you would invite if she were a man', which presumably included professors of archaeology but not night club hostesses, although we never found out exactly where the line might be drawn.

But there was some doubt about whether the university would still be open then. Many men and a few women were leaving for the forces or war work, and after the fall of France and the evacuation of the British army from Dunkirk, the future looked bleak. In the summer, Franklin helped her mother with her work for refugees and pondered possibilities outside Cambridge. But after another hiking holiday (this time in the Pennines), she was able to pick up her studies, just as the Blitz on London was starting. This forced the Franklin family to leave for the safer quarters of a rented house in Radlett, in Hertfordshire.

In her third year, Franklin was supervised by Fred Dainton (later Lord Dainton), who described her as having 'very high personal and scientific standards' and being 'uncompromisingly honest'.* But her most important new acquaintance was Adrienne Weill, a French-Jewish scientist, born in 1903, who had been one of Marie Curie's students; she had fled from France ahead of the Nazis and was now working at the Cavendish Laboratory. Her family were acquaintances of Franklin's great-uncle Herbert. In her final (Part Two) exams, Franklin 'only' got a second, rather than the first that might have been expected.† The problem was that she had poor exam technique, and insisted on answering each question as thoroughly as possible before moving on to the next. As a result, in two of her papers she only answered two questions instead of the three that were required, so it was impossible to score more than 66 per cent on those papers. This thoroughness may also have played a negative part later in her career, as we shall see. In spite of this, with her track record and Dainton's recommendation, she was awarded a further scholarship of £15 and a grant from the

* Quoted by Anne Sayre, *Rosalind Franklin & DNA*, Norton, New York, 1975.

† This did not, of course, entitle her to a degree, since this was still before Cambridge allowed women such status.

Department of Scientific and Industrial Research to stay on for a year to carry out a research project.

This first experience of research was not a success. For most of the time, she lived in a rented room on Mill Road, but late in 1942 she moved in with a group of French refugees that were sharing a house with Adrienne Weill, where she revelled in their company and enjoyed speaking French, which seems to have released her from some of her stiff-upper-lip British reserve. Around the same time, she found a job with the British Coal Utilisation Research Association (BCURA) at a lab near Kingston-upon-Thames. This was rewarding work in its own right, studying the behaviour of different kinds of coal and charcoal under different conditions, and directly relevant to the war effort because of the importance of coal as a fuel for industry. It was the kind of job that would never have been offered to a woman had it not been wartime.

After a few months, Franklin moved to a house on Putney Common, which she shared with her cousin Irene and a friend. She took on most of the housekeeping duties and also became a volunteer air raid warden. As ever, holidays were spent hiking in the hills. At the end of 1943, however, she was back living with her parents, as Irene had got married. As the war ended and the job with BCURA finished, in 1945 Franklin briefly went back to Cambridge to write up her research on coal for a thesis which saw her awarded a PhD by the university later that year – it was, as we have seen, OK for a woman to have a PhD, even if she could not technically have a BA. When she presented a paper on this work to the Royal Institution in 1946, the audience included a French crystallographer, Marcel Mathieu, who had been advised by Weill (now back in Paris) to check Franklin out while he was in London. At his recommendation, she was soon offered a dream job, working at the Laboratoire Central des Services Chimiques de l'Etat, in Paris (now the Institut National de Recherche Chimique Appliquée). Naturally, she accepted.

In Paris, where she arrived in February 1947, Franklin blossomed both professionally and socially. It was here that she learned the techniques of X-ray crystallography, in particular applied to the study of coal and clay. Her salary was equivalent to about £5 per week, barely sufficient to cover living expenses, but she was free to follow up whatever aspects of the research she found interesting; for all practical purposes, she was her own boss. In a letter home,* she wrote: 'Of course my standard of living is lower than at home … but I find infinite kindness and goodwill among the people I work with. All this is far more important than a large meat ration or more frequent baths.'

Franklin's superior in Paris was Jacques Mering, born in 1904, who had a distinguished career and in 1959 became director of research at the Centre National de Recherche Scientifique (CNRS), but whose fate was to be remembered today primarily as the man who taught Rosalind Franklin X-ray crystallography. Unfortunately, even though the dangers of working with X-rays were well known by the late 1940s, there were few safety precautions at the lab, and even these were sometimes ignored. As a result, on at least one occasion, the badge Franklin wore to monitor her exposure exceeded the safety limit, and she had to stay away for several weeks. On a brighter note, in 1948 the University of Cambridge recognised the error of its ways (well, one of its ways), and retroactively awarded degrees to all the female students such as Franklin who had passed the necessary examinations in previous years.

In 1949, Franklin, with a growing reputation as a crystallographer, was starting to think about returning to England, although she was determined not to do so until she found a job there. She applied for a post at Birkbeck College, where J.D. Bernal had

* Quoted by Maddox.

Rosalind Franklin, 1950

returned from his sojourn in Cambridge, but was turned down – as was another applicant, Francis Crick, who would instead end up at the Cavendish Laboratory in Cambridge. In March the following year, not entirely sure whether to leave Paris and considering her options, Franklin visited King's College in London to discuss possibilities with John Randall, the head of their biophysics unit. The upshot was that she was awarded a three-year research fellowship, funded by Imperial Chemical Industries (ICI) and worth £750 a year, to work in Randall's group. The initial intention was for her to use her skill as an X-ray crystallographer to study the structure of proteins. But instead of taking this up at the start of the academic year, she postponed her move to London until January 1951. The delay produced a significant change in the work assigned to her.

In the summer of 1950, Maurice Wilkins, a biophysicist working at King's on the study of DNA, obtained some particularly pure samples from a Swiss colleague, Rudolf Signer, and Randall decided that this should be the initial focus of Franklin's work. He wrote* to tell her that:

> It now seems that it would be a good deal more important for you to investigate the structure of certain biological fibres … as far as the experimental X-ray effort is concerned there will be at the moment only yourself and [Raymond] Gosling† … Gosling, working with Wilkins, has already found that fibres of desoxyribose‡ nucleic acid derived from material provided by Professor Signer of Berne give remarkably good fibre diagrams.

* Quoted by Robert Olby, in *The Path to the Double Helix*, Macmillan, London, 1974.

† A research student working for his PhD at King's.

‡ The technically correct chemical name, although now more often shortened to deoxyribose.

So Franklin arrived at King's in January 1951 with the clear understanding that she would be the only person, assisted by Gosling, working on the X-ray investigation of DNA. But Randall had not consulted Wilkins about this, and he was understandably aggrieved to see a newcomer given this plum task using 'his' DNA.

The atmosphere at King's was very different from Paris – among other things, there were two dining rooms, one for use by both men and women, the other for men only – and almost immediately Franklin regretted her decision to leave Paris. The situation was not helped by being set against Wilkins, through no fault of her own. Randall's solution to the overlap of their interests, for reasons known only to himself, was to assign the latest equipment, purchased with her in mind, to Franklin, and to give her the Signer DNA, while letting Wilkins and his team work with older equipment and less pure DNA. This did not make for a happy working atmosphere.

It was soon after Franklin arrived at King's that James Watson and Francis Crick, working at the Cavendish Laboratory in Cambridge, learned of the progress being made in DNA studies. In May 1951, Wilkins gave a talk describing the progress so far at a conference in Naples, where Watson was in the audience. Back in Cambridge, Watson persuaded Crick, with whom he shared an office, to join him in trying to work out the structure of the molecule. All they needed, Watson suggested, was an understanding of the rules of chemistry and some of the data being gathered by the crystallographers. They didn't actually have to do any experiments themselves. They were both supposed to be working on other projects, but Crick agreed. The story of how events then unfolded has been told many times,* and the details need not concern us here.

On 21 November 1951, Franklin gave a talk at King's summing up her work so far. The lecture notes, which have survived, include

* We recommend the Brenda Maddox version.

the comment: 'The results suggest a helical structure (which must be very closely packed) containing 2, 3 or 4 co-axial nucleic acid chains.' Watson was present in the audience.

By then, in spite of the excellent progress being made with DNA, Franklin had decided that she had to get away from King's. In January 1952 she took a holiday in Paris and asked if there was any chance of returning to the Laboratoire Central; when the answer was 'no', on her return to London she visited Bernal at Birkbeck College to check out the possibilities there. He was more positive, although no definite plans were made for a move at that time. So in 1952, while she continued to work with Gosling on DNA, Franklin was already planning her next move.

The key to Franklin's steady progress over the next few months can be seen in her approach to her final exams – just as she had felt compelled to answer each question thoroughly before moving on to the next, instead of leaping from the idea of a helix to trying to guess the detailed structure of the molecule, she wanted to check every possibility, dotting all the i's and crossing all the t's before publishing her conclusions. She said to a colleague: 'We are not going to speculate, we are going to wait, we are going to let the spots on this photograph tell us what the [DNA] structure is.'* Among other things, as part of this process she took some of her photographs to Oxford for Dorothy Hodgkin to see. Hodgkin's response was that they were the best examples of their kind that she had ever seen.†

With everything going well and the end of the DNA project in sight, Franklin's thoughts turned to the future. In June 1952 she told Randall of her wish to move to Birkbeck, and in July her application

---

* See Samuel Schindler, 'Model, Theory, and Evidence in the Discovery of the DNA Structure', *British Journal for the Philosophy of Science,* volume 59, page 619, 2008.

† Quoted by Maddox.

to spend the third year of her fellowship there was approved, with the move originally planned to take effect from 1 January 1953, although in the end it was postponed until March, partly because Franklin suffered a severe bout of flu and was unable to work for a month. This gave comfortable time for her and Gosling to complete their project and write up the results for publication. But while they were unhurriedly finishing their work, Watson and Crick were scrambling to beat everyone to the determination of the structure of DNA, by hook or by crook.

The Cavendish team had access to all the important results from King's – Wilkins visited them to discuss progress, and Watson visited King's, where, unbeknown to Franklin, he was even shown one of the key photographs that had so impressed Hodgkin. They also 'got hold of'* the annual report Randall gave the Medical Research Council, one of his major sponsors. In addition, unhampered by the need to do any experiments themselves, they had time to eagerly read up on everything they could find on the chemical properties of DNA.

In February 1953, Franklin completed writing up two long papers on DNA for the journal *Acta Crystallographer*, and in her notebooks stated that DNA is a 'two-chain' helix. She also drafted a short paper, including Gosling's name, reporting this conclusion. On 7 March, the day after the two long papers reached *Acta Crystallographer*, Watson and Crick finished building their famous model of the double helix, and invited Wilkins to come and see it, which he did on 12 March. When he learned they were planning to publish in *Nature*, he suggested that he could send in a short paper to back up their work. He did not mention Franklin, but when Gosling heard the news he told Wilkins that he and Franklin had already drafted a short paper on DNA. The result was three short papers in the 25 April 1953 issue of *Nature*. First came Crick and Watson, claiming that their model was based on the chemistry and

* Maddox's expression.

mentioning that the X-ray data supported their conclusion.* Then came Wilkins's team, discussing the X-ray data in general terms. Third came Franklin and Gosling, with a slightly modified version of their paper presenting the direct evidence for the double helix.

It's an intriguing thought that without Franklin's dose of influenza the original version of this paper could have been finished by the beginning of February, and might have appeared in *Nature* before Crick and Watson completed their model building. Regardless of this might-have-been, however, it would have been more accurate for the Franklin–Gosling paper to come first, with Crick and Watson acknowledging that it was their work which pointed the way to the detailed model of the double helix, although the Cavendish team had added the crucial chemical ingredients which show how the two strands of the helix twine together.

In 1962, the Nobel Prize for Physiology or Medicine was awarded to Crick, Watson and Wilkins for this work. Franklin missed out, not because she was a woman, but because by then she was dead.† Had she lived, the Nobel Committee would have faced a ticklish problem since their rules say that a maximum of three people can share one Prize. But had she lived, as we shall see, Franklin might well have won another Nobel Prize in her own right.

By the time the three *Nature* papers appeared, Franklin was already settling in at Birkbeck, where the company was more congenial, but the working conditions were appalling. With many buildings damaged or destroyed in the Blitz, her X-ray equipment was in the

* In a much longer paper published in the *Proceedings of the Royal Society* in 1954 (volume 223, page 80), Crick and Watson said that without Franklin's data 'the formulation of our structure would have been most unlikely, if not impossible'.

† Gosling also missed out, but until very recently the contributions of research students were often ignored by the Nobel Committee.

basement of what had originally been a row of terraced houses in Bloomsbury, while her office was in a room on the fourth floor. Her status was peculiar – as a researcher with outside funding, she was not technically a faculty member, and at first she was refused admission to the faculty dining room. There were other niggles. When her original research grant ran out, the college requested funding from the Agricultural Research Council to support her and a team of assistants for three years. The appropriate salary for someone with her status should have been £1,100 per year, but the ARC offered £1,080, apparently just to assert who was in charge, which she had no choice but to accept. What mattered was her work.

The reason for the ARC's interest was that Bernal had steered Franklin towards determining the structure of the tobacco mosaic virus, TMV, which, as its name suggests, causes a disease in tobacco plants. Thanks to people like Franklin, we now know that viruses are made of pieces of DNA or RNA* (the sister to DNA), wrapped in protein molecules. They work by attacking cells and injecting the nucleic acid into the cell, where it hijacks the cell's machinery and makes many copies of the virus, which burst out of the cell and infect other cells. Because of their complicated composition, determining the structure of TMV and other viruses was much harder than determining the structure of DNA.

When she started this work, Franklin was still best known to the outside world for her research on coal. As a result, in 1954 she was invited to participate in a major research conference on coal in the United States, and was able to combine this with visits to labs working on viruses, including Woods Hole (where Watson was based at that time) and Caltech. She found what she called 'the carbon crowd' boring and uninspired, while the biologists were 'first class' and 'impressive'. By the time she returned to England, she was known as

* Ribosenucleic acid, the one that is in TMV.

a member of the second group. In the October 1954 issue of *Scientific American* there was an article by Francis Crick describing his work with Watson, in which he said that they 'were convinced that we could get somewhere near the DNA structure by building scale models based on the X-ray pattern obtained by Wilkins, Franklin and their coworkers'. At that point the key collaborator who would help her to an even higher level came on the scene.

Aaron Klug was born in Lithuania in 1926, but emigrated with his family to South Africa when he was two. He had just completed a PhD in physics at Cambridge University when he arrived at Birkbeck at the end of 1953, where he was fascinated by Franklin's work and soon joined her team. They made an ideal pairing. Franklin produced superb X-ray photographs, Klug provided the theoretical framework to interpret the images. Before long, they were joined by two PhD students, John Finch and Kenneth Holmes, who both went on to become Fellows of the Royal Society. Franklin was also a frequent visitor to Cambridge, where she discussed viruses with Crick. She was by 1955 the leader of a team doing first-class work which was recognised as such by the biophysics community. Their research then covered a whole slew of viruses, including those infecting potatoes, turnips and peas, as well as TMV. The only people who didn't seem to appreciate her work were her sponsors, the ARC.

When Franklin requested that her status be raised from senior scientific investigator to principal scientific investigator to reflect the real situation, the ARC refused. When university pay scales were partially revised to (gradually!) bring women up to the same level as men, her salary should have increased to £1,250 a year. Again, the ARC refused, then grudgingly increased it to £1,200, once more deliberately setting it lower than her entitlement to show her who was boss. The only bright spot on that front was that Klug was, after much lobbying, put on the ARC payroll. But with this funding due to end in 1957, Franklin had to start thinking about the future.

In 1955 and 1956 Franklin was in great demand as a speaker at scientific conferences – usually the only woman present apart from wives – and scientific papers were pouring out from her group. When she was invited back to the United States for another extended visit, partly funded by the Rockefeller Foundation, there was no mention of coal. This time, she was the head of virus research at Birkbeck College, a recognised leader in the field. The American connection also opened up a new possibility – a colleague suggested she apply for funding from the US National Institutes of Health (NIH). Also while she was in America, Franklin received a request from Lawrence Bragg, the director of the Royal Institution, for two large models, each about five feet tall, based on her analyses of viruses to be made for the Science Hall of the Brussels World's Fair (Expo 58) opening in 1958. But at the peak of her career the first symptoms of Franklin's terminal illness appeared when she experienced severe abdominal pain while in California. Soon after her return to England at the end of August 1956, she was diagnosed with cancer.

On 4 September, Franklin was operated on and had a tumour and one ovary removed. A month later, a second operation removed the other ovary and her womb. After convalescing with friends and family, optimistic that the problem had been solved, she returned to work at Birkbeck, initially only for half days, where her first priority was to apply to the NIH for funds. The ARC had renewed its funding (£4,300) for one more year, but emphasised that this must end in March 1958. Soon after receiving this news, at the end of April, Franklin was back in hospital and told that there was no longer any hope of a cure.

Without telling anyone there how serious the situation was, she again returned to Birkbeck, where in July she had news that the NIH had approved an award of £10,000 per year (equivalent to more than a quarter of a million pounds today) for three years – the largest grant Birkbeck had ever received. Against the advice of her doctors, Franklin attended two conferences in Europe later that year, combining these with a driving holiday through Italy with her sister Jenifer. And she started a new research project, investigating the structure of the polio virus. But in November she began receiving chemotherapy at the Royal Marsden Hospital.

Yet again, between bouts of treatment she returned to work, where she was appointed as a research associate in biophysics, with a salary of £1,700 a year, to take effect in October 1958. Best of all, she was invited to use the NIH funding to move her team to Cambridge, where a new molecular biology building was under construction. Her scientific legacy was assured. But she never lived to see any of this. Rosalind Franklin died on 16 April 1958, the day before her virus models went on public display in Brussels. She was still only 37.

Franklin's team was left in good hands. Klug took over,* and when they moved to Cambridge in the early 1960s, the Medical Research Council took over their funding. Building on Franklin's legacy, Klug would receive the Nobel Prize in 1982 'for his development of crystallographic electron microscopy and his structural elucidation of biologically important nucleic acid-protein complexes', including TMV. In his Nobel lecture, he said of Franklin that 'had her life not been cut tragically short, she might well have stood in this place on an earlier occasion' – or, indeed, alongside him. He was, of course, not referring to her work on DNA, but to her finest work, on viruses. The 'structural elucidation of biologically important nucleic acid-protein complexes', which she pioneered, was greatly helped in the years following her death by the development of electronic computers, which were pioneered by another researcher working in Cambridge at the same time as Klug. Her name was Lucy Slater, and she was just two years younger than Franklin.

* With Finch and Holmes he completed the work on the structure of the polio virus and published the results in June 1959.

CHAPTER

11

Lucy Slater (5 January 1922–6 June 2008)

Pioneered the development of computer operating systems

Computers today are so user-friendly that you don't have to have any programming skills or even be able to type to use them. You can give them instructions by moving a pointer around on a screen, or simply by speaking to them. But how does the computer know what you want it to do? The average computer user doesn't know what is going on inside the machine any more than the average car user knows what is going on under the bonnet. But there has to be a system that translates the wishes of the user into a language the computer can understand – and translates the answer back. These systems were pioneered in the 1950s and 1960s, alongside the development of the hardware on which computers run. And one of the unsung pioneers of this work was a woman who survived being blown up during the Blitz in Portsmouth to rise through the ranks of a male-dominated profession and come within an ace of winning a Nobel Prize.

Lucy Slater's life brings us into the twenty-first century, but it began in England in 1922, six years before women were granted equal voting rights with men. Unlike some of her predecessors, there was never any objection to her choice of career. Her parents were both graduates of the University of Manchester – they met before the First World War and married in 1916. Her mother, an arts graduate, was an active suffragette who became a teacher during the war, while her father, who became a distinguished chemist, did war work on explosives, which led to a career as a scientific adviser to the Royal Navy and a move to Parkstone, near Bournemouth, where Lucy was born and had a comfortable childhood, with an unusual early education. Her mother, who was related to the pioneering chemist John Dalton, an early promoter of the atomic theory, was not working at the time, and Lucy did not go to school but was allowed to largely self-educate, using the books lying around the house. This triggered an early enthusiasm for mathematics.

Her life changed dramatically in 1931, when she was nine. Her father, who was a keen ballroom dancer, collapsed during a competition and was dead before he arrived at the hospital. Her mother had to go back to work, teaching at the girls' grammar school in Portsmouth, but Lucy's formal education in mathematics began a little later, when she was about sixteen, taking courses at Portsmouth technical college – she never attended school in the usual way. Lucy also studied music and became a proficient pianist, playing in a local jazz band in her teens. This led to a particularly memorable evening the night before the outbreak of the Second World War.

On 2 September 1939, seventeen-year-old Lucy went to see the great saxophonist Coleman Hawkins, who was playing with the British trumpeter Nat Gonella at a club in Southsea, near Portsmouth. Their pianist was a Swede who was eager to get home in view of the rapidly changing political situation and left immediately after the gig. But the jazzers, as jazzers always do, wanted to go on to another club for a more informal session. Lucy was sufficiently well known locally to get plucked out of the audience and roped in on piano. The next day, war was declared; and on 10 January 1941 the little jazz club was blown to pieces in an air raid, fortunately with nobody inside.

As soon as she turned eighteen, in January 1940, Lucy joined the organisation known as the ARP (from Air Raid Precautions), where she worked initially answering telephones and later as a van driver delivering supplies to various ARP posts, while continuing her studies to prepare for university and still managing to fit in piano lessons. In this wartime organisation men and women had equal status and responsibilities; Lucy's duties included being out during air raids to, among other things, spot incendiary bombs and smother them with sand before they could do much damage. In the autumn of 1940 she was cycling home when a low-flying Messerschmitt fired a burst of machine-gun bullets her way. Fortunately, it missed. A little later she had another near miss with death, but this time the work for her university entrance exams saved her life.

Together with a group of friends, Lucy was in the habit of visiting a jazz club on Fridays, where she would often play the piano. On 10 January 1941, she decided, on impulse, to go home and revise for the exam she would be taking the following Monday, leaving a group of her friends to go off together to the club. It was about 4pm, and very soon what turned out to be one of the biggest air raids on Portsmouth of the entire war began. As the sirens went off, the group of jazz enthusiasts hurried into a nearby air-raid shelter. It received a direct hit from a bomb, and most of them were killed. It was the same raid that destroyed the club where Lucy had played with Coleman Hawkins. She lost two boyfriends during the war. One was killed in the Blitz, the second, a flyer with the Fleet Air Arm, was lost from a carrier escorting a convoy to Murmansk. They had planned to marry after the war.

The Slaters' house in Portsmouth was also damaged in the Blitz, and they moved to Southampton, where Lucy's mother taught Latin and Greek at Southampton University College, as it then was. Lucy entered the college in October 1942, and claimed that she only passed the entrance examination because the authorities were lenient on students whose studies had suffered because of air raids. Southampton College was not then a university in its own right, and their degrees were awarded by the University of London. Under wartime conditions, the degree course was crammed into two years of intense study, but Lucy continued with her ARP duties, driving supplies (mostly at night) from Southampton to Portsmouth. After graduating in 1944, she began work as a teacher in Winchester High School, although still living in Southampton, and obtained a teaching diploma in 1945; but the war had not quite finished with her.

In June 1944, Southampton was undergoing a second Blitz, this time from the V1 'flying bombs', an early kind of cruise missile powered by a primitive jet engine. Lucy was on ARP duty and about to go down the stairs leading to a bunker when one of these missiles approached, and she stopped to watch it go by. As she watched, the engine cut out and the bomb nosed over towards the ground. As she hurriedly turned to go down, she stumbled and the bomb exploded nearby. She was blown down the stairs, suffering a head injury that left a permanent dent in her skull, a broken wrist and ankle, and damage to her back that was never diagnosed, as she didn't bother going to hospital, although she did see a doctor about the wrist and ankle. The various injuries healed with minimum attention, although for years she consumed large quantities of over-the-counter drugs such as aspirin to alleviate the pain from her back – a souvenir of what turned out to be one of the last raids on Southampton, as the Allied forces invading France soon overran the V1 launch sites.

While teaching in Winchester, Lucy was also recruited to teach trigonometry to trainee radar operators at an outfit called the Signals Research and Development Establishment (SRDE) near Chichester. Somebody in authority thought that it would be a good idea for the radar operators to understand the geometry of a situation where radar-directed anti-aircraft guns were being fired at their targets; Lucy thought the whole idea was ridiculous, but every Friday she would travel to the SRDE to give a lecture on the subject. At the end of the war, Lucy got job at Southampton University College, teaching a refresher course in mathematics to demobilised troops who wanted to renew their interrupted university studies. But within a few months, she picked up the threads of her own university career.

Her first degree had been a so-called 'ordinary' degree, in maths and music, but in 1945 she obtained a grant to resume her studies at Southampton and upgrade it to an honours degree in mathematics, awarded in 1947. The next opportunity came about because although she had completed her honours degree in two years, she had received a grant covering the usual three-year course – so she had a year's worth of financial support to use up. She went to Bedford College in London, where she worked on solutions to the Schrödinger equation describing the nature of the hydrogen atom, and was awarded an MA in 1949. This was cutting-edge research at the time, not least because of its relevance to the theory of the hydrogen bomb, and her talent led to the award of a further grant to develop her work into a PhD, awarded in 1951. The next step in Dr Slater's career involved her first brush with blatant sexism.

Around the time she was finishing her London PhD, in 1951, Slater visited the Festival of Britain, a post-war celebration of the arts and science held in London's South Bank area. By chance, she bumped into Eric Mutch, whom she had met at the SRDE. He told her about the plan to build an electronic computer in Cambridge, where he was now working, which piqued her interest. But before she could follow up that interest, she received an offer she chose to refuse.

Out of the blue, Slater received an invitation to be interviewed for a job at Fort Halsted, which was the location of the UK's secret atomic weapons research project (this attempted head-hunting gives you some idea of the importance of her work and the high esteem in which she was held at the age of 29). The invitation did not specify the work involved, but she had a pretty good idea; she went to Harwell, home to the civilian and not secret UK Atomic Energy Research Establishment (AERE), for the interview. But although the existence of AERE was not secret, security was still tight, so Fort Halsted would be an even less comfortable environment in which to work. Slater was therefore already inclined to reject the opportunity when she received the offer of a job with the rank of senior scientific officer, but for which, because she was a woman, she would be paid the salary of a scientific officer.

You can imagine the reaction of her mother, the former suffragette; Lucy, equally outraged, replied that she would not consider the post unless she was paid the same as a man. When the authorities replied that on second thoughts they would indeed pay her the appropriate salary, she took great delight in turning them down anyway, and moving on to Cambridge.

This move had also been complicated by bureaucracy, although not because she was a woman. Since 1949, Slater had spent a lot of time in Cambridge, collaborating with fellow mathematicians and specifically finding out about their pioneering work on computing. She had also visited Oxford, where she was disappointed to find nobody working in the area, and Manchester, where there was another team (which included Alan Turing) developing an electronic computer. But she had no official position in Cambridge, partly because at the time Cambridge had a rather superior attitude to other universities and regarded somebody with 'only' an external degree from London as a kind of second-class academic citizen. The usual route for such outsiders to be accepted on a Cambridge PhD programme involved a preliminary course roughly equivalent to an MSc elsewhere (this is still the situation in most cases). Possession of a 'proper' PhD from the University of London elevated her in their eyes sufficiently to remove this requirement, and once again, thanks to her ability and capacity for hard work, she had completed her thesis at Bedford College with a year left on her latest grant. So she was able to transfer to Girton College in Cambridge, taking her grant with her. Girton was then a college only for women (or 'ladies', in the language of the day); Girton only admitted men in 1979, and even then it was one of the first mixed-sex colleges.

Slater's arrival at Girton brought about an encounter with sexism that ran completely counter to her wartime experiences of equality. New students at Cambridge were required to participate in an induction ceremony known as matriculation, even if, like 29-year-old Lucy, they had already acquired a couple of degrees. For the matriculation ceremony, the ladies were required to wear a black outfit with white gloves, with no jewellery, and to assemble together before proceeding to the Senate House for the ceremony. Lucy had neither a black outfit nor white gloves, although she did possess a navy-blue suit with matching shoes. Clothes were still rationed in Britain in 1951, and even if she had been able to afford new clothes, she did not have the necessary ration coupons. The authorities at Girton said that wearing the wrong colour would be letting the college down, so she gave them an ultimatum. Either she would join the procession in her dark-blue outfit and dark gloves, or she wouldn't go at all. They agreed to let her do her own thing, and nobody noticed. In 1968, when she became a Doctor of Science (Sc.D.), the highest degree awarded by the university and an extremely rare honour for a woman in those days, she took great delight in wearing a much brighter outfit than was usual for the ceremony.

In another sign of the times, in the 1950s Girton had a house in Cambridge (the college itself is a little out of town) where the ladies could meet to socialise and take refreshments without running the risk of mingling with male students in the regular cafes (let alone, heaven forbid, meeting men who were not students). A ludicrous precaution for someone who had worked nights in the ARP and had been unofficially engaged.

The time at Girton led to a Cambridge PhD, awarded in 1953, and the award of a Bateson research fellowship, at Newnham College, from 1953 to 1956. It was while at Newnham that, alongside her official job developing programming tools, Slater started doing work on applied economics and published a series of theoretical papers that earned her a D.Litt (Doctor of Letters, a kind of super-PhD) from London – but that lay in the future when she started at the computer lab in 1951.

The computer being developed in Cambridge was called the Electronic Delay Storage Automatic Calculator (EDSAC), and it became the first stored-program computer in regular use, as a practical tool rather than an experimental test bed. The 'stored program' bit of its name means that once instructions were fed in to the machine, it could store them in its memory while carrying out a particular task, then that program could be replaced by another to perform a different task. Slater's initial work involved developing the systems that enabled the computer to work at all (systems analysis), and programs that would then enable EDSAC to solve problems such as the Schrödinger equation – what we now call coding. A computer system is the code that communicates between the programs written by human beings and the nuts and bolts of the hardware. If you ask a computer to find, say, the square root of two, the system translates your request into strings of binary code, ones and zeroes that run through the electronics. Without an operating system, a computer is considerably less useful, although an expert programmer can write code which will run directly on what they call 'bare metal'.

Operating systems have two basic functions: to make the machine work more effectively and to make the machine easier to operate. And since the operating system itself requires some of the memory of the computer to run, the less memory you have, the harder it is to write an operating system. EDSAC had a total memory of 512 35-bit words, amounting to just under 18,000 bits (a bit is basically an on/off switch). In modern jargon, memory is measured in bytes, where each byte is 8 bits, so EDSAC had roughly 2,500 bytes, or 2.5k, of memory. Today, the smallest amount of memory you will find in a tiny Raspberry Pi home computer takes up a couple of gigabytes – roughly a million times the entire memory of EDSAC. Slater's systems work was pioneering, and it laid the foundations for much of the development of computing.

The working conditions in the lab were far from ideal. The engineers worked on the computer during the daytime, so the programs had to be run at night, in a room at the top of a four-storey building. The temperamental machine needed constant attendance, shared by two operators, one who worked up until midnight and the other taking over until morning. One of the first problems Slater encountered was that there was no toilet for women in the building, even though there were many female employees in non-academic jobs there in the daytime. The nearest toilet for women was across a courtyard and down a flight of steps into a basement – an obstacle course that might have to be traversed at night, in the rain, while leaving EDSAC to its own devices. Slater, who'd had no problem sharing the toilet with men in her ARP days, was having none of that, and solved the problem by singing whenever she was in the toilet as a warning to any men who might approach.

Apart from personal convenience, leaving EDSAC to its own devices was not a good idea. The heart of the machine was a huge array of electric valves (known as tubes in America), which glowed red and put out a lot of heat – Slater said as much as 100 kilowatts. Many of these components came from an unlikely source – the US Army Air Force. After the war, the USAAF was shutting up shop and moving home from many of the airbases in East Anglia, not far from Cambridge. They would have what amounted to a garage sale of things they didn't want, such as surplus radar equipment. The Cambridge computer pioneers would snap up anything that looked interesting and worry about what to do with it when they got it back to the lab.

The heat when the computer was running sometimes caused the valves to come loose in their sockets, halting the program. In such circumstances, the standard solution was to give the machine a kick to get it going again; a run of a quarter of an hour without interruption was good going in the early days. The power running through a somewhat haphazard array of cables and experimental bits and pieces tacked on to the machine in the daytime by the engineers also posed a fire hazard, which needed monitoring. But there was an upside. The room was nice and warm, and whoever was running the machine in the night could pick up food from a nearby pub before closing time and keep it hot on top of the computer until it was time to eat.

Conditions in the crowded offices below the computer were also less than ideal, and Slater had to share a desk with the astronomer Fred Hoyle; fortunately, they got on well. Hoyle became an enthusiastic promoter of the use of computers in solving astrophysical problems, and in the 1960s obtained the best civilian computer in the UK for his Institute of Theoretical Astronomy.*

* The many users of this machine, an IBM 360-44, included JG when he was a research student.

Lucy Slater, 1959

With permission from the Mistress and Fellows, Girton College, Cambridge.

As the only academic woman in an essentially all-male environment (there were, of course, female secretaries and other support staff), Slater did make one concession. She wore long skirts and deliberately drab clothing to avoid any possibility of her colleagues showing anything other than a professional interest in her – she kept such relationships away from work. The other problem with men, according to Slater, was that they didn't have the patience for the kind of work involved in getting programs to run in those days. She was often approached by colleagues from other departments for help and was happy to spend weeks debugging the code and getting it to run correctly, especially since this sometimes involved 'unofficial' payments that were not declared to the income tax authorities. These payments were not always in cash. Slater used to tell the story of when, a few years later, she did some semi-official work for the manufacturers of Mars bars, who paid with a crate of assorted chocolate goodies.

This was after she had a fall, towards the end of her Newnham Fellowship, and was unable to get up. Her back had finally had enough, and she spent weeks in plaster in Addenbrooke's Hospital getting it straightened out. This mostly resolved the problem, as long as she was careful to avoid standing for long periods. Fortunately, programming (writing algorithms) is a sitting-down job.

Slater continued this work, paid out of a series of short-term funds, until 1962. Her interests increasingly turned to economics, especially after the advent of EDSAC II in 1958. To give you some idea of the step up this represented, the work involved a process known as inverting a matrix. EDSAC could invert a three-by-three matrix, but EDSAC II could invert a twelve-by-twelve matrix, although it would take a whole afternoon to do it. Even if you know nothing about matrices or inversion, this was clearly a leap forward.

The key influence in Slater's official move into economics was Allan Brown, who was the number two in the economics department at the time. He was one of the people who lacked the patience to do his own coding and had pushed work her way on a semi-official basis to start with. But when EDSAC II became available, he got the necessary approval from the university to set up an official economics 'computing unit' (actually just Slater and one assistant) to work on the application of computing for developing models of the economy. Her economics guru in the project was a former director of the Department of Applied Economics (DAE), Richard Stone (later Sir Richard), who received the Nobel Prize in 1984 for his work on developing a model to track economic activity. This owed more than a little to Slater's work.

Slater's own formal move to the DAE, as assistant director of research, came in 1962, when she was 40. She stayed there until she retired in 1982; in the 1960s, one of us (MG) worked for her in a very junior capacity, and experienced firsthand her generous encouragement of young women. Many senior economists, including the future governor of the Bank of England, Mervyn King, came to the DAE to learn how to apply computers in their work, but Slater was just as keen for her young colleagues to benefit from her knowledge. Although she looked somewhat intimidating at first sight – she still wore dowdy clothes and had a dress sense like that of the TV detective Vera – she would perch on the edge of a desk to rest her bad back and chat about everything from the choice of biscuits for the tea break to the political situation. It took me quite a while to realise that she was indeed 'the boss'. If there were any high jinks going on in the room where we worked and footsteps were heard coming down the hall, we would all silently bend over our work waiting for the door to open. If it was Slater who entered, there would be a sigh of 'it's only Lucy' and normal service would be resumed. This was all highly unusual in the rigidly hierarchical DAE (indeed, Cambridge) of the time. But Slater had no respect for authority even when she was the authority. She even offered to teach me programming, but unfortunately in spite of her ideas about the patience of women I had no aptitude for the task.

In the mid-to-late 1960s, Slater's work had a direct influence on government economic policy. Nicholas Kaldor, later Baron Kaldor, was one of the leading Cambridge economists and became the principal advisor on economic matters to the government of Harold Wilson. Much of the data he used came from the computer work at the DAE, in spite of the fact that his skill at mathematics was limited (according to Slater) to working out sums on the back of an envelope.

Before and after her appointment at the DAE, Slater was also roped in as a consultant, officially or unofficially, on various other computer projects, and was involved in some capacity in the development of almost all the early British machines. Her experience of setting up computer systems was second to none, and her expertise was widely recognised. But it wasn't all plain sailing.

When the University of Essex was establishing their computing facility, they invited Slater, who was due for a sabbatical from Cambridge, to advise them how to proceed, and she went there for six months. At that time, there was a major British company called International Computers Limited (ICL), which was supplying equipment to Essex. In a fit of generosity, they asked Essex if they needed anything in particular, as a donation. Slater identified their most pressing need as a storage unit (basically a memory bank) that would normally cost £10,000, an enormous sum in those days. ICL seem to have felt initially that this was an attempt by a mere woman who obviously didn't know what she was talking about to take advantage of their generosity. When she was invited to visit ICL and explain her reasons for asking for the storage unit, however, she made such an impressive case that they gave the university twice what she had originally asked.

Another bizarre incident occurred during the early days at the DAE. In those days, computer programs were written as holes punched in a long paper tape,* and somebody – Slater never found out who – stole one of the tapes, the original algorithms of the economic model, planning to use it themselves. It would have been useless without the specialist knowledge of Slater and her colleagues, but after this Slater would always put a deliberate fault in to her programs and write a little routine to correct the fault, which she kept as a paper tape in her pocket. Without her and her override tape, nothing would work – an early counter to hacking.

Around this time, the physical computational work was transferred to new machines at the University of London, and it was becoming less important for people like Slater to be present where the programs were run. Even before the advent of satellite links, data could be transferred across the Atlantic using undersea telecommunications cables, and there was a major UK/USA collaboration, funded by the Chase Manhattan Bank, which involved shuffling data between London, New York, San Francisco and Tokyo.

After the excitement of the pioneering early years, Slater's work became more or less routine in the run-up to 1982, when she officially retired but remained with the DAE as a research associate and consultant. In retirement she developed her interest in genealogy, particularly the history of her mother's Dalton family line. And in 1984, when Stone was awarded his Nobel Prize, for 'having made fundamental contributions to the development of systems of national accounts and hence, greatly improved the basis for empirical economic analysis', the DAE held a big party at which he referred to her as his dea ex machina (the goddess of the machine) and said she deserved a share of the award. Some of her colleagues have suggested that, as in many other cases, the Nobel Committee was again showing a prejudice against women.

* My job at DAE was essentially punching the holes in the tape.

But awards never interested Slater. When asked why she stuck with computers for so long, she had a ready explanation:

> They used to turn me on. It's rather difficult to explain, but if you're tired and you've been working all night, and then the computer suddenly springs to life, and the program works, and it produces what you want it to do – it's rather like driving a fast car! If you've ever had a really nice car, and you've driven it fast: I used to have a Riley with two carburettors, and its top speed was supposed to be ninety, but I could get it up to a hundred and twenty. Why I didn't kill myself I don't know! And why I didn't get run in – of course, there weren't many police on the roads in those days; it was about 1960. That feeling. It gave me a kick. And I suppose it's a kind of sexual thing, because the kick that you get out of trying to make love was the same sort of kick I was getting out of making the computer work. It seems awful to say you were in love with a computer! But it was that kind of heightened sensibility, a feeling that something had happened that was very important. Well, in all probability it wasn't, you know, but it was to me. And that's the only way I can express it: that running a fast, really big system, like those CDCs,* and getting the satellite link to work all around the world, knowing it was the first time it had been done – it turned me on! That's the only phrase for it.†

* Control Data Corporation was a major US computer manufacturer in the 1960s.

† Lucy Slater, an oral history conducted 9 April 2001 by Janet Abbate, IEEE History Center, Piscataway, NJ, U.S.A. https://ethw.org/Oral-History:Lucy_Slater#About_Lucy_Slater. The IEEE History Center has a collection of more than 800 oral histories in electrical and computer technology which can be accessed via http://ethw.org/Oral-History:List_of_all_Oral_Histories.

Perhaps the love affair with computers was a compensation for the way all the men she loved seem to have vanished from her life. As well as her father's early death, there were the two serious boyfriends she lost during the war, and a later lover who died of a heart attack. Lucy Slater herself died of heart disease, on 4 June 2008 at Addenbrooke's Hospital in Cambridge. But she left some advice to young women based on her own experiences:

> It seems essential to me that when a woman gets married, she looks on it as a temporary situation and not a for-life situation that's going to provide her with a rain check. That would be my advice. Because even today, men step under buses, and have heart attacks, and run off with other women. You can't trust a man: that's the sum total of my experience!*

* Op.cit.

CHAPTER 12

Vera Rubin
(23 July 1928–25 December 2016)

Discovered the widespread existence of dark matter in the Universe

Lucy Slater epitomises the kind of successful female scientist who lives for her work, does not marry and has no family commitments. This does not seem to have been a deliberate choice for Slater – she might well have married and had children if the opportunity had not been snatched away from her. But would that have prevented her from having a productive career? The example of her near-contemporary Vera Rubin shows that even in the 1950s and 1960s it was possible for a woman to combine a successful family life with a successful career. Vera had a long and happy marriage, raised four children who all became scientists, and carried out astronomical observations which underpin our modern understanding of the nature of the Universe. Along the way she campaigned for women in science and broke through several glass ceilings. Hers is an extreme example; we can't all make scientific discoveries of that importance, and she certainly needed a lot of support from her family, including a man who didn't step under a bus, have a heart attack or run off with other women. But the fact that it is possible at all refutes the idea that women must always choose between a family and a career, and provides a suitably optimistic ending for our book.

Vera's parents came from immigrant Jewish families. Her father was born in Lithuania in 1897, and was originally Pesach Kobcheski, but became Peter Cooper after his family moved to America in the early 1900s. He was eight at the time. Vera's mother, Rose Applebaum, was born in Philadelphia in 1900, but her own parents had also emigrated from what was then the Russian Empire. She married Peter in 1924, four years after he completed a degree in electrical engineering; their first daughter, Ruth, was born in 1926, and Vera in 1928.

Like many young families, the Coopers' lives were badly affected by the economic depression of the 1930s, and in 1934 they had to move in with Peter's mother, after her husband died. As the economic situation improved, after a succession of short-term jobs, including the proverbial insurance salesman, in 1939 Peter Cooper found employment using his electrical engineering skills with the Department of Agriculture in Washington, DC. Vera became a pupil at the H.D. Cooke Elementary School, where she encountered two contrasts with school in Philadelphia – one good, one bad. The girls' previous school had had an old-fashioned attitude to teaching and discipline, so strict that the left-handed Vera had been forced to write with her right hand. The new school was much more relaxed, and learning was a pleasure. But unlike the Philadelphia school, it was segregated; there were no pupils of colour there, and segregation also applied to the seating on the buses.

It was the move to Washington that opened Vera's eyes to the fascination of astronomy. She shared a large double bed with her sister Ruth, and the bed was pushed up against a window. As the elder sister, Ruth claimed the side away from the window, which was easier to get in and out of. Vera had the window, which faced north, and through which she could see the stars, with very little glare from the feeble street-lighting of the time. Gazing out each night, she noticed that the stars seemed to move across the sky as the Earth rotates, and that the patterns change with the seasons. She became hooked on the heavens.

In 1941, the year Vera moved on to Paul Junior High School, her father was transferred to the US Army Signal Corps as part of their preparation for what was clearly a coming war; he was already there at the time of the Japanese attack on Pearl Harbor on 7 December that year, and he stayed with them until the war ended in 1945. During these years in high school, Vera read everything she could lay her hands on about astronomy, and even built her own telescope out of a long cardboard tube and a couple of lenses, the larger one about three inches across. She also went to meetings of an amateur astronomers' club, where she heard talks by eminent astronomers such as Harlow Shapley, one of the first people to realise that our Milky Way is just one 'island' in space among a multitude of other galaxies.

She sailed through junior high, and on to the Calvin Coolidge Senior High School, where her love of science received a jolt from the physics teacher, whose class she later described as a 'macho boys' club' and in which she was made to feel unwelcome. On the first day in his class, he described to his pupils the two kinds of scientist. One kind, he said, made discoveries through insight and brilliance. The examples he gave were all men. The other kind, he said, made discoveries by hard work and diligence. The only example he gave was Marie Curie. He succeeded over the months that followed in putting Vera off physics for life, although she excelled at mathematics, where she had the benefit of 'the best teacher I met in my entire career', and also enjoyed anything that involved making things. In 1945, the year the war ended, she graduated from high school, turned seventeen and won a scholarship to Vassar College.

When she told her physics teacher about the scholarship, he replied: 'You should do OK as long as you stay away from science.'

Vassar was the natural place for an aspiring female astronomer to study. The college had been founded for female students in 1865, and appointed as its first professor of astronomy was Maria Mitchell, who had made her name by discovering a comet in 1848. When Mitchell retired in 1888, she was succeeded by her former pupil Mary Whitney, who was in turn succeeded by one of her own students, Caroline Furness, and then, in 1936, by Maud Makemson. So there was an unbroken line of female astronomers in charge of the observatory at Vassar up until the arrival of Vera in 1945.

During the war, Vassar, like many universities, had introduced a first degree course which packed all the usual work into three years, instead of four, in order to produce graduates more quickly. In 1945, new students were offered a choice of taking their degree over three or four years; Vera, characteristically, chose the shorter option, which meant more intense work but would produce quicker results. As well as her formal courses, she became a paid assistant at the observatory and learned about the practical side of astronomy; over the next three summers she also worked at the US Naval Research Laboratory in Washington on a variety of experimental projects, gaining more practical experience. One of the highlights of her time at Vassar was a visit to the college from the charismatic Richard Feynman, from Cornell University; their paths would soon cross again, partly as a result of a dramatic change in Vera's life in 1947.

During her first year at Vassar, Vera's parents had moved to an apartment in southeast Washington, where they became friendly with another Jewish family, Ben and Bess Rubin. When Rose Cooper and Bess Rubin discovered that the Coopers' daughter was just two years younger than the Rubins' son Bob, they hatched a plan to introduce their offspring to each other, which they did over a family dinner in the early summer of 1947. The matchmaking – if that was what it was – succeeded beyond their expectations. Vera and Bob fell deeply in love, and after he returned to Cornell, where he was completing his first degree and planning to study for a PhD, and she went back to Vassar, Bob took every opportunity to make the relatively short journey to visit her at weekends. At Thanksgiving that year, less than five months after they had met, they told their parents they were engaged. Vera also applied for a graduate studentship at Cornell in order to be with Bob.

Maud Makemson gave Vera a glowing reference,* saying that: 'I rely on her as I would on a graduate assistant, and I have indeed found her to be more intelligent, quicker, more independent and more reliable than some graduate students I have had … it is a joy to have her around the observatory as she is always happy and good-natured …'

In spite of the high praise, at the time Makemson was disappointed with Vera's decision to get married. Makemson made the natural assumption, for the time, that marriage and children would bring an end to the career of her most promising pupil. But she had underestimated Vera – and would not be the last to do so.

Vera graduated in May 1948, and became Vera Rubin when she married Bob on 25 June that year, a month before her twentieth birthday. Bob had graduated just nine days earlier, having been through college on a navy programme. He had enlisted in 1944, when he was eighteen, but the war ended before he saw active service. He was now about to begin his studies for a PhD in physics at Cornell, where Vera had, of course, been accepted as a graduate student for a Master's degree in astronomy. Her thesis adviser was Martha Stahr, only eight years older than Vera, who was the first

* Quoted by Mitton & Mitton.

female member of the Cornell Faculty of Arts and Sciences. In physics, both Vera and Bob were supervised by Feynman, which fortunately meant that Bob could help her with the work she found most difficult.

At that time, it was only a little over twenty years since it had been realised that the Milky Way is just one galaxy among many, and very little was known about the Universe at large. For her thesis, Vera analysed the way galaxies move through space, using published measurements of the spectra of 108 galaxies. Although the project did not discover anything significant, spectroscopy would be at the heart of Rubin's later work.* It depends on the fact that different elements – such as hydrogen, oxygen and carbon – produce distinctive patterns of lines in the rainbow spectrum of light produced when light from an object is passed through a prism. When the object moves towards us, the lines are seen shifted towards the blue end of the spectrum (a blueshift), and when the object is moving away from us they are shifted towards the red (a redshift). The amount of the shift tells us how fast the object (such as a galaxy) is moving relative to us.†

Although Rubin's thesis was not spectacular, her overall adviser, William Shaw, surprised her by saying that it was good enough for the results to be presented at the next meeting of the American Astronomical Society (AAS), in December 1949. But, he said, as Rubin was expecting a baby, due in the autumn, she would, of

* For clarity, now she is an astronomer we will refer to her as 'Rubin', and her husband as 'Bob'.

† This is related to, but not the same as, the famous cosmological redshift which indicates that the Universe is expanding. The cosmological effect is caused by space stretching; the velocity or Doppler effect, named after the physicist Christian Doppler, is caused by objects moving through space. The two effects have to be disentangled in the kind of work she was doing.

course, not be able to go to the meeting and he would make the presentation – in his own name! When Rubin insisted that she would be able to make the presentation herself, he had no choice but to agree.

The subsequent logistical nightmare highlights Rubin's attitude to combining her family and professional lives. She passed her Master's examination on 2 October, and her son David was born on 28 November. The meeting was on 28 December, but by chance it would be at Haverford College, just ten miles from the home of her grandmother. Rubin's mother and aunt would look after the baby at her grandmother's house while the 22-year-old mother made her first professional presentation as an astronomer. For the time being, though, it would be her last. Family came first while Bob finished his PhD, awarded in 1951, and started work at the Applied Physics Laboratory (APL) near Washington, associated with Johns Hopkins University. But, bored witless by the daily routine of a young housewife, Rubin almost immediately began searching for a way to earn her own PhD, and found a home at Georgetown University.

Georgetown was a college founded by the Jesuits, where the obliging Father Francis Heyden, head of the small astronomy department, not only agreed to take her on, starting in February 1952, but already had a programme geared to students who could only attend lectures in the evenings. Her time as a 'housewife' had lasted for just six months. Bob's salary paid her fees, her mother babysat every Tuesday and Thursday evening while Rubin attended lectures, and in the daytime and long into the night she combined her duties as a mother with her research. Those duties soon included Judith, born on 15 September 1952.

The result of all this, what she later described as the hardest time in her life, was a PhD, awarded in 1954, just before her twenty-sixth birthday. The topic, like her Master's thesis, had no long-term importance; the most significant impact on Rubin during that time came from the opportunity to attend a scientific meeting at Ann Arbor in Michigan in the summer of 1953. There, she met eminent astronomers including Walter Baade, Margaret Burbidge and her husband Geoffrey Burbidge, and had discussions with George Gamow, whom she already knew, as he worked in Washington and had given her advice about her thesis work. It also gave them the opportunity to notice this unusually impressive young woman.

Although Rubin joined the American Astronomical Society (AAS) and kept up with developments by reading the scientific journals, for a time she gave the family priority, and worked part-time as a teacher at a local college. But in 1955, Father Heyden offered her a job as his research assistant on a project analysing solar eclipse data. It wasn't a particularly interesting project for her, but it was her first professional employment as an astronomer. The fact that Bob got a job at the University of Illinois in Urbana-Champaign didn't matter, as the eclipse work could be done from home, and she took the birth of a third child, Karl, in January 1956, in her stride. But in 1957, Bob moved on to a post with the National Bureau of Standards, and the family was back in Washington. There, in January 1958, Rubin received a letter inviting her to succeed Maud Makemson as professor of astronomy at Vassar. It was impossible for her to accept – one of the very few times that she put family before her career. Instead, she continued to work at Georgetown, now on a project analysing the spectrum of the Sun. The skills she developed in spectroscopy would eventually lead her to her breakthrough work on galaxies, and, serendipitously, around the same time she had an opportunity to take practical steps to use those skills.

In 1959 the French astronomer Gérard de Vaucouleurs, a friend and colleague of the Burbidges, moved to Harvard College Observatory. His special interest was the distribution of galaxies across the Universe, and he had heard about her presentation at the AAS meeting in December 1949. He wrote to her in 1958 asking for a copy of her Master's thesis, and in 1959, after an exchange of letters, offered to help her find work in this area. In the spring that year, de Vaucouleurs attended a meeting at the National Academy of Sciences in Washington, and they met face to face. The upshot was that Rubin obtained funds for a two-year project analysing his data, starting in June 1960. She would also be teaching a course at Georgetown on the structure of the Milky Way. And just to make sure she didn't have any idle time, the Rubins' fourth child, Allan, was born on 23 May 1960. None of this stopped her attending a two-week summer school in the Netherlands on 'The structure and evolution of the galactic system', where she was brought up to date 'on astronomical discoveries I had overlooked in the 1950s while getting my Ph.D. degree and raising four children'. She also renewed acquaintance there with Margaret Burbidge and her husband, who were investigating the way galaxies rotate and trying to determine their masses. The mass can be worked out by measuring how fast the stars in a galaxy are moving around its centre – the faster they are moving, the more mass there must be to hold them in orbit.

The work Rubin did for de Vaucouleurs up until the end of her association with the project in 1964 involved analysing images of bright galaxies; it did not produce any dramatic results but, like the solar spectrum studies, honed Rubin's skills at studying the light from galaxies. Alongside this work, however, Rubin was developing her own line of research, initially involving some of her students at Georgetown. This would be the first direct step on the road to her major contribution to astronomy, and it involved measuring the rotation of galaxies – starting with our own Milky Way.

Before this work, astronomers had made the seemingly natural assumption that stars further out from the galactic centre orbit around the centre more slowly than stars closer in, just as the outer planets of our Solar System (such as Saturn) circle the Sun more slowly than the inner planets (such as Venus), in accordance with Newton's laws. Rubin made no assumptions, but simply set out with her students to analyse, using published data on spectroscopy and other astronomical evidence, how fast stars at different distances from the centre of the Milky Way are moving (their rotational velocity). The data were buried in huge star catalogues, from which the students had to dig out suitable stars for study, but the result was unambiguous. From the central nucleus of the Milky Way at least as far out as the orbit of the Sun, everything rotates at the same speed.

Rubin sent the paper describing these surprising results to the *Astronomical Journal*. She had included the names of all the students involved as co-authors, but the editor objected to this. So she told him she would publish it elsewhere, and he changed his mind. It duly appeared in print in October 1962 with all the authors named,* and the conclusion that 'the decrease in rotational velocity expected for Keplerian orbits is not found'. Around the time she was doing this work, Rubin attended a major astronomical meeting – a Symposium of the International Astronomical Union – where, although she was able to meet once again the top people in her field, the unconscious sexism of science was revealed. The official announcement of the Symposium mentioned that alongside the scientific work going on, 'a program is being prepared for the ladies'. The assumption was that the only women present would be the wives of the astronomers. This in spite of the fact that one of the main contributors to the meeting was Margaret Burbidge,

* Vera C. Rubin, Jaylee Burley, Ahmad Kiasatpoor, Benny Klock, Gerald Pease, Erich Rutsheidt, and Clayton Smith, *Astronomical Journal*, volume 67, page 491, 1962.

from whom Rubin learned about the importance of determining rotation curves for other galaxies.

Rubin now made the key change in her career that would lead her to her most important discovery. She became, gradually, an observer, using the telescopes to obtain her own data. This started when a visiting astronomer, Martin McCarthy, spent the 1962–63 academic year at Georgetown and taught her the necessary skills using the relatively modest telescope there. Just as she was becoming an observer, Bob was awarded a fellowship which meant that he could work at the University of California, San Diego, for the year 1963–64. This was the base where the Burbidges worked, and Rubin, backed with a powerful recommendation from McCarthy, was appointed as a visitor to work with them, using the 36-inch telescope* at the Kitt Peak National Observatory. The result was a series of nine published papers, co-authored with the Burbidges, which established her reputation as a skilled observer.

In 1964, Rubin attended another IAU Symposium, in Hamburg, this time as an 'invited participant'. One of the most important people there was Allan Sandage, who was the most senior observer using the big optical telescopes at the Mount Wilson and Palomar Observatories in California. He had taken over the role from Edwin Hubble, who had proved that many of the fuzzy patches of light on the sky are indeed galaxies beyond the Milky Way. Sandage was a good friend of the Burbidges and had already met Rubin. At a social gathering in Hamburg, he casually asked her if she would like to use one of the big telescopes. This was a bombshell. As late as 1964, women were still officially not allowed to use these telescopes; one reason cited for this, Lucy Slater would surely have been amused to hear, being that there was only one toilet. It was also officially

* The number refers to the diameter of the main mirror in the telescope.

thought that the male night assistants who helped to operate the telescopes would refuse to take orders from a woman.

In fact, there had been one woman who unofficially got access to the telescopes in the 1950s. The Burbidges made such a good team partly because Margaret was an observer and Geoffrey was a theorist. Thanks to this combination, although Rubin was the first woman officially allowed to observe on the mountain, she was not the first woman to use the telescopes. Margaret Burbidge got access to them with the collusion of Allan Sandage, who persuaded the director of the Observatories, Ira Bowen, to turn a blind eye. Margaret told me (JG) about the circumstances when I interviewed her for the journal *Nature* in 1972, shortly after her appointment as the first female director of the Royal Greenwich Observatory (RGO), then based at Herstmonceux, in Sussex.

There was an intriguing subplot to this appointment. Until that time, the posts of director of the RGO and Astronomer Royal had always been linked. But Margaret was not appointed as Astronomer Royal, and the title, from then on purely honorary, went to a man, the radio astronomer Martin Ryle; there has never been a female Astronomer Royal. It is possible that this was not a result of sexist discrimination, but it naturally led me to ask Margaret if she had experienced any problems with sexism during her career. She replied with the details of her semi-official observing.

In the mid-1950s, Geoff, although a theorist, applied for time at the telescopes on Mount Wilson, and got a temporary appointment there. Margaret, as his wife, was allowed to accompany him, and even to 'help' him with the observing, but because the living quarters on the mountain only had accommodation for men (it was known as 'the monastery'), they would have to stay in a separate building on the site, which Margaret described as a 'log cabin'. Ten years later, in 1965, Rubin was not only awarded five nights' observing time at Mount Palomar, but even allowed to stay in the monastery, where the two-bedroom top floor of the three-storey building was reserved for her. And the night assistants were, of course, happy to work with her.

With this incentive, Rubin gave up her post at Georgetown, where her teaching commitments had left little time for research, and managed to secure a niche at the Carnegie Institution's Department of Terrestrial Magnetism (DTM) in Washington – their first female researcher.* She was paid two-thirds of the usual salary, allowing her to leave the office at 3pm to pick up the children from school. But that didn't stop her working at home in the evening. Her idea of 'part time' was more like time and a half. The release from administrative and teaching duties into full-time research was, however, such a joy that the family noticed a lightening of her mood. A few weeks after she started at the DTM, Allan, then five years old, asked how much she had to pay them to let her work there.

The original aim of the DTM had indeed been to map the Earth's magnetic field, but it had long since diversified into other areas of research, including astronomy. One of the most exciting projects there at the time was the work of Kent Ford, developing so-called image tube technology for astronomical spectroscopy, in the form of a spectrograph. In a nutshell, the image tubes would make it possible to get spectra of much fainter objects than could be done using traditional photography. Rubin chose to share an office with Ford, whom she planned to work with. It was their collaboration that produced her most famous work.

By the mid-1960s, Rubin was established as an observer, in a job where she could focus on research while still raising four children. And she had both the tools and a collaborator that could provide

* The Carnegie Institution was also responsible for the Mount Wilson and Palomar Observatories, but the two centres operated entirely independently.

her with detailed spectra of very faint objects. The pair worked at various telescopes throughout their careers together, but here we shall concentrate on their findings, not how they found them.

Their first big success came from analysing the light from the newly discovered objects called quasars, which we now know are the energetic cores of distant galaxies. In the second half of the 1960s there was a mad scramble by observers to measure the redshifts in the light from these objects.* Rubin and Ford got caught up in the scramble, and published measurements for several quasars. This demonstrated the power of the new spectrograph; in one case they obtained a spectrum in twenty minutes using a 69-inch telescope, where traditional techniques required a six-hour exposure using a 200-inch telescope. But both of them disliked the competitive nature of this work, where teams raced to measure the biggest known redshift, or the greatest number of redshifts, and they soon left this area to pick up the threads of Rubin's work on galaxy rotation. Before, she had measured the way stars orbit around the Milky Way; now, with Ford's help and the image-tube spectrograph, she would measure the rotation of our neighbouring galaxy, M31 in Andromeda.

Although the Andromeda galaxy is a near neighbour in cosmological terms, it is still too far away for Rubin to be able to track its rotation by measuring velocities of individual stars. Instead, she studied bright blobs of hydrogen gas, called HII regions, which lie among the stars. Radio astronomers had already made measurements of the rotation by studying cold hydrogen gas in the Andromeda galaxy, and had found the same kind of flat rotation curve that Rubin had found for the Milky Way. But many other astronomers had been reluctant to take this evidence at face value,

* These are the cosmological redshifts that indicate distances across the Universe, not Doppler redshifts that indicate motion through space.

Vera Rubin, 1974

NOIRLab/NSF/AURA, CC BY 4.0, via Wikimedia Commons

not trusting the radio observations, and were waiting to see what observations using visible light would tell them. After two years of painstaking work in collaboration with Ford, in December 1968 Rubin was able to announce at a meeting of the AAS that the radio astronomers were right. The rotation curve of the Andromeda galaxy is indeed flat. Which meant that there must be a lot of unseen dark matter, way out from the central bright region of the galaxy, exerting its gravitational influence on the visible stars. And if this was true for both the Milky Way and the Andromeda galaxy, might it be true for all galaxies?

For those who knew their astronomical history, this would not have come entirely as a surprise – but in 1968 hardly anyone was familiar with the relevant bit of history. The term 'dark matter' was not new to astronomy at the end of the 1960s. It had actually been first used in an astronomical context by the Swiss-born astronomer Fritz Zwicky, back in 1933. He had been investigating the way galaxies move within clusters, associations in which the individual galaxies are moving in different directions, like the bees in a swarm, while the whole cluster is carried along by the expanding Universe. He found that the Doppler velocities of these galaxies are so large that the clusters ought to fly apart, like the fragments from an exploding bomb. The only way to stop this happening is if some unseen material is holding the galaxies in its gravitational grip. He referred, in a paper written in German, to the 'surprising result that dark matter* is present in much greater amount than luminous matter'. But Zwicky's conclusion had largely been ignored for more than 30 years, as had suggestions along similar lines made from time to time by other astronomers.

* In German, 'dunkle Materie'.

It would soon be impossible to ignore what Rubin discovered. But before she got her teeth into that problem, she returned to a theme related to her early work on galaxies. Together with Ford, she obtained redshifts of nearly a hundred galaxies in an attempt to find out if the Universe is expanding evenly in all directions (isotropically), or if there are local deviations from this behaviour. Their data suggested that there is a local deviation from this 'Hubble flow', but it later turned out that the way distances to the galaxies used in their sample had been measured was flawed, so their conclusion did not stand up.

So it wasn't until 1976, the year of her forty-eighth birthday, that Rubin, working with Ford and other collaborators, began an intense study of the rotation curves of galaxies, a project which occupied her for the next ten years.

The Milky Way and the Andromeda galaxies are both flattened, disc-shaped star systems, obviously rotating. Like many disc galaxies, they have bright spiral arms, streams of bright stars twining around the disc. These flattened systems are often referred to as 'spiral galaxies', although 'disc galaxy' is a better term, since not all disc galaxies have spiral arms. There are also other kinds of galaxy, but for rotation measurements disc galaxies are ideal, especially when they lie edge on, or nearly edge on, to our telescopes. Then, by using a narrow slit cutting vertically across the image of the galaxy, it is possible to take spectra at different distances out from the centre and measure the Doppler velocities, after subtracting out the cosmological redshift, with one side of the disc rotating away from us and the other towards us. The light being analysed doesn't come from individual stars, but is the combined light from all the stars at a particular distance from the centre of the galaxy. This is simple in principle but very hard in practice. The image tubes developed by Ford enabled Rubin to probe very faint regions of these galaxies, much further out than any previous studies had reached. They also had the benefit of large new telescopes that became available in the mid-1970s, in both the northern and southern hemispheres. There was, by this time, no difficulty in Rubin, with an established reputation as an observer, being allocated time on these instruments.

By 1976, the problem of measuring rotation curves in this way was becoming pressing, because the radio astronomers kept finding flat rotation curves, implying the presence of dark matter, from their studies of radio emission from hydrogen gas in disc galaxies. There is no sense in which Rubin 'discovered' dark matter. But it was her work, essentially over the period from 1976 to 1986, that provided the clinching evidence that visible galaxies are surrounded by spherical halos of dark matter, with at least ten times more dark matter than the mass of all the bright stars put together.

In 1980, about halfway through the project, she wrote in a major paper in the *Astrophysical Journal** that 'The conclusion is inescapable that non-luminous matter exists beyond the optical galaxy' in all the examples they had studied. This was the key paper that made everyone sit up and take notice; it has now been cited in papers by other scientists more than a thousand times, a phenomenal record. By the end of 1985, Rubin had measured the rotation curves of almost a hundred disc galaxies; they were all flat. Dark matter is now an integral part of our understanding of the Universe at large, and of the way galaxies form and evolve.

Hardly surprisingly, Rubin, who was 58 in 1986, did not make any further major contributions to astronomy. But throughout her career, alongside her research, as soon as she had any influence at all she used it to promote the place of women in science, and to encourage young female astronomers. In the early 1970s she

* Volume 238, pages 471–87.

was part of a working group appointed by the AAS to report on the status of women in astronomy; the report appeared in 1974, and documented problems such as the pay gap between men and women doing the same work, concluding that 'women face far greater obstacles in almost all aspects of their professional careers'. The next year, a personal example highlighted this. Rubin's daughter Judith started her studies for a PhD in astronomy at the University of Minnesota; when the authorities learned that she was about to get married, they advised her to give up the course since she could not possibly have both a successful marriage and a career in astronomy. She ignored their advice. In due course, Judith became an astronomer working at the University of Massachusetts Amherst; her brother David worked with the US Geological Survey, Karl became a mathematician at Ohio State University and Allan became a geologist at Stanford.

One of Rubin's pet hates was the exclusive use of masculine pronouns and examples in formal documents. When the journal *Physics Today* published an article with advice on how to give lectures, not only were all the lecturers assumed to be 'he', but the one female example in the piece was a dancer who fell on her face. 'What much of the scientific community appears to be saying to young women is that it is male,' she wrote in her letter of complaint. But progress was being made. The same year, 1982, Margaret Burbidge became the first female president of the American Association for the Advancement of Science, and more women than ever were gaining access to the big telescopes. In her book *Bright Galaxies, Dark Matter*, Rubin mentions her delight at receiving a cable from a team at the Cerro Tololo Observatory in South America, telling her that for the first time all the observers there that night were women. In 1986, the astronomer Wendy Freedman, who would later become the director of the Carnegie Observatories, wrote to Rubin saying: 'Thanks for speaking out. I often get the feeling that if things are any easier for women like me, it's because women like you have taken the time and energy to make it so.'*

Rubin herself continued to be an active observer into the twenty-first century, but increasingly devoted less time to the telescopes and more to her grandchildren. She also kept a desk at the DTM where she could analyse the data. But after Bob died in January 2008, following a long illness suffering from multiple myeloma, she made only one more observing trip, to the Lowell Observatory. The death of Judith in 2014, from the same cancer that had killed Bob, seems to have triggered a decline in Rubin's mental faculties, and she died on 25 December 2016. One of her essays from *Bright Galaxies, Dark Matter* sums up her philosophy, and provides a suitable conclusion to our book:

1. There is no problem in science that can be solved by a man that cannot be solved by a woman.
2. Worldwide, half the brains are in women.
3. We all need permission to do science, but for some reasons, that are deeply ingrained in history, this permission is more often given to men than women.

* Quoted by Mitton & Mitton.

SOURCES AND FURTHER READING

Chapter 1

Eunice Foote, 'Circumstances affecting the heat of the Sun's rays: Art. XXXI', *The American Journal of Science and Arts*, 2nd Series, v. XXII/no. LXVI, November 1856, pages 382–83: https://ia800802.us.archive.org/4/items/mobot31753002152491/mobot31753002152491.pdf

Roland Jackson, 'Eunice Foote, John Tyndall and a Question of Priority', *Notes and Records of the Royal Society*, volume 74, 2020, pages 105–18: https://royalsocietypublishing.org/doi/pdf/10.1098/rsnr.2018.0066

Elizabeth Wagner Reed, *American Women in Science before the Civil War* (PDF). Minneapolis, Minnesota: University of Minnesota. OCLC 28126164. (Originally published 1992): https://web.archive.org/web/20160708160623/http:/catherinecreed.com/assets/writings/women_in_science.pdf

JohnTyndall, 'Note on the transmission of heat through gaseous bodies', *Proceedings Royal Society of London*, v. 10, 1859, pages 37–39.

Chapter 2

Otto Hahn (translated by Willy Ley), *A Scientific Autobiography*, MacGibbon & Kee, London, 1967.

Lise Meitner, 'Looking Back', *Bulletin of the Atomic Scientists*, volume 20, 1964, pages 2–7.

Ruth Lewin Sime, *Lise Meitner*, University of California Press, Berkeley, 1996.

Mike Sutton, 'Hahn, Meitner and the discovery of nuclear fission', Chemistry World, 5 November 2018: https://www.chemistryworld.com/features/hahn-meitner-and-the-discovery-of-nuclear-fission/3009604.article

Chapter 3

N. Byers, 'The Life and Times of Emmy Noether: Contributions of E. Noether to particle physics', arXiv:hep-th/9411110, November 15, 1994: https://arxiv.org/abs/hep-th/9411110

A. Dick, *Emmy Noether, 1882–1935*, Birkhäuser, Boston, Mass., 1981.

Dwight Neuenschwander, *Emmy Noether's Wonderful Theorem*, Johns Hopkins UP, updated edition, 2017.

David Rowe, Mechthild Koreuber, *Proving it Her Way: Emmy Noether, a Life in Mathematics*, Springer, 2020.

Chapter 4

Bruce Bolt, Inge Lehmann, *Biographical Memoirs of the Royal Society*, volume 43, 1997, page 286.

Inge Lehmann, 'Seismology in the days of old', *EOS*, volume 68, 1987, page 33.

Lif Lund Jacobsen, 'Intellectually gifted but inherently fragile – society's view of female scientists as experienced by seismologist Inge Lehmann up to 1930', *History of Geophysical and Space Sciences*, volume 13, 2022, pages 83–92: https://doi.org/10.5194/hgss-13-83-2022

M.W. Rossiter, *Women Scientists in America*, Johns Hopkins UP, Baltimore, 1984.

Chapter 5

Denis Brian, *The Curies*, Wiley, Hoboken, New Jersey, 2005.

M.-H. Chiu, Penny Gilmer, D.F. Treagust, 'Celebrating the 100th Anniversary of Madame Marie Sklodowska Curie's Nobel Prize in Chemistry', *Sense*, Rotterdam, 2011.

Eve Curie (translated by Vincent Sheean), *Madame Curie*, Pocket Books, New York, 1964.

Irène Joliot-Curie, Nobel Lecture 1935: https://www.nobelprize.org/prizes/chemistry/1935/joliot-curie/lecture/

Sharon McGrayne, *Nobel Prize Women in Science*, second edition, Carol Publishing, Secaucus, New Jersey, 1998.

Chapter 6

Interview of Cecilia Payne-Gaposchkin by Owen Gingerich on 5 March 1968, Niels Bohr Library & Archives, American Institute of Physics, College Park, MD USA: www.aip.org/history-programs/niels-bohr-library/oral-histories/4620

John Gribbin, *Stardust*, Penguin, London, 2001.

Katherine Haramundanis (editor), *Cecilia Payne-Gaposchkin: An Autobiography and Other Recollections*, Cambridge University Press, 1984. (Haramundanis is the married name of Cecilia's daughter.)

Donovan Moore, *What Stars Are Made Of*, Harvard UP, 2020.

A.G. Davis Philip & Rebecca A. Koopman (editors), *The Starry Universe: The Cecilia Payne-Gaposchkin Centenary*, Davis Press, Schenectady, 2000.

Chapter 7

Nina V. Federoff, 'Barbara McClintock', National Academies Press, Washington, DC, 1995.

John Gribbin and Mary Gribbin, *On the Origin of Evolution*, Collins, London, 2020.

Evelyn Fox Keller, *A Feeling for the Organism*, Freeman, San Francisco, 1983.

Barbara McClintock, Nobel Lecture 1983: https://www.nobelprize.org/uploads/2018/06/mcclintock-lecture.pdf

Chapter 8

Georgina Ferry, *Dorothy Crowfoot Hodgkin: A Life*, Granta, London, 1998. Revised edition, Bloomsbury, London, 2019.

Dorothy Crowfoot Hodgkin, Nobel Lecture 1964: https://www.nobelprize.org/uploads/2018/06/hodgkin-lecture-1.pdf

G. Dodson, J. Glusker, S. Ramaseshan, K. Venkatesan, *The Collected Works of Dorothy Hodgkin*, Interline Publishing, Bangalore, 1994.

Horace Freeland Judson, *The Eighth Day of Creation*, Cold Spring Harbor Press, New York, Enlarged edition, 1996.

Chapter 9

Noemie Benczer-Koller, 'Chien-Shiung Wu 1912–1997: A Biographical Memoir', National Academy of Sciences, Washington, DC, 2009.

Tsai-Chien Chiang (translated by Wong Tang-Fong), *Madame Chien-Shiung Wu: The First Lady of*

Physics Research, World Scientific, Singapore, 2014.

Martin Gardner, *The New Ambidextrous Universe: Symmetry and Asymmetry from*

Mirror Reflections to Superstrings, Courier Corporation, Chelmsford, 2005.

John Gribbin, *Six Impossible Things*, Icon, London, 2019.

Chapter 10

Brenda Maddox, *Rosalind Franklin*, HarperCollins, London, 2002.

Matthew Cobb, *Life's Greatest Secret*, Profile, London, 2016.

Horace Freeland Judson, *The Eighth Day of Creation*, expanded edition, Cold Spring Harbor Press, 1996.

James Watson, Alexander Gann, Jan Witkowski, *The Double Helix*, expanded annotated edition, Simon & Schuster, New York, 2012.

Chapter 11

Girton College Archive, Personal Papers of Lucy Joan Slater: https://arcspace-pub.lib.cam.ac.uk/repositories/19/archival_objects/370800

Mary Gribbin, unpublished recollections.

Lucy Joan Slater, 'EDSAC: Recollections on early days in the Cambridge Computing Laboratory', *Journal of Economic and Social Measurement*, volume 29, number 1, January 2004

Lucy Slater: an oral history conducted in 2012 by Janet Abbate, IEEE History Center, Piscataway, NJ, USA.

Chapter 12

Interviews conducted by David DeVorkin for the American Institute of Physics in 1995 and 1996: www.aip.org/history-programs/niels-bohr-library/oral-histories/5920-1 and www.aip.org/history-programs/niels-bohr-library/oral-histories/5920-2

Jacqueline Mitton and Simon Mitton, *Vera Rubin: A Life*, Belknap Press, Cambridge, Mass., 2021.

Vera Rubin, *Bright Galaxies, Dark Matter*, American Institute of Physics, Woodbury, NY, 1997.

Ashley Jean Yeager, *Bright Galaxies, Dark Matter, and Beyond*, MIT Press, Cambridge, Mass., 2021.

INDEX

Page numbers in **bold** refer to images

abolition of slavery 2
Academy of Sciences, France 68–9, 80, 83
Ackermann–Teubner Prize 52
Acta Crystallographer 171
Adams Prize 91
Adler, Hanna 56
Agricultural Research Council (ARC) 173, 174, 175
Alembic Club, Oxford 127
Alvarez, Luis 145
American Association for the Advancement of Science (AAAS) 3, 10, 12, 213
American Astronomical Society (AAS) 200, 202, 203, 210, 212
American Geophysical Union 15
American Journal of Science and Arts 3
American Physical Society 157
Anderson, Carl 78
Annual of Scientific Discovery, The 10
Applied Physics Laboratory (APL) 201
Archimedeans 162
ARP (Air Raid Precautions) 179–80, 184
artificial radioactivity 67
Artin, Emil 52
Aspect, Alain 151
Association of German Chemists 23–4
Association of Scientific Workers 162
Astronomer Royal 206
Astronomical Journal 204
astronomy 88, 90–6, 99–100, 196–7, 200–13
Astrophysical Journal 212
atomic bomb 82, 145, 147
Atomic Energy Commission, France 84
Atomic Energy Research Establishment (AERE) 182
atomic physics 87–8, 91–2
Avery, Oswald 114

Baade, Walter 202
Balfour Declaration 1917 160
Balfour, Arthur 160
Beadle, George 107, 113–14
Becker, Herbert 76
Becquerel 14
Bedford College, London 181, 183
Bell, John 52
Berkeley, University of California 144–6, 148
Berlin 19–20
Bernal, J.D. 126, 128–32, 134, 135, 137, 166, 170, 173
Big Bang theory 154
Biot, Jean-Baptiste 13
Birkbeck College, London 166, 170–4, 175
Blundell, Tom 139
Bohr, Niels 33, 34, 36, 37, 56, 57, 79, 87, 92, 126
Bolt, Bruce 65
Boltzmann, Henriette 18
Boltzmann, Ludwig 18
Bothe, Walther 76
Bowen, Ira 206
Bragg, Lawrence 124, 137, 175
Bragg, William 123–4, 128, 145, 162
Branly, Edouard 69
British Astronomical Association (BAA) 89
British Coal Utilisation Research Association (BCURA) 165
British School of Archaeology 124
Britten, Benjamin 139
Brookhaven National Laboratory 148
Brown, Allan 188–9
Brussels World's Fair (Expo 58) 175, 176
Bryn Mawr College, USA 53–4
Burbidge, Geoffrey 202, 203, 205–6
Burbidge, Margaret 202, 203, 204, 205–6, 213

California Institute of Technology 144
California, University of 137, 205

Caltech 173
Calvin Coolidge Senior High School 197
Cambridge, University of 25, 26, 36, 38, 56–8, 86–9, 94, 125, 128, 130, 162, 166, 174, 176, 181–2, 189–90
Carnegie Institution of Washington 113, 119, 207–8, 214
Carnegie Observatories 213
Cavendish Laboratory, Cambridge 36, 57, 87, 164, 168, 170–2
Centre National de Recherche Scientifique (CNRS) 166
Cerro Tololo Observatory 213
Chadwick, James 25–6, 77, 80, 84
Chamberlain, Neville 162
Chase Manhattan Bank 191
chemistry 20, 22, 25–8, 29–36, 69, 76–80, 122–3, 126–7, 130–40, 161–2
China 140, 141–3, 146, 147–8, 151, 156, 157
Chinese Cultural Revolution 156
Churchill, Winston 139
Clauser, John 151
coal 165–6, 173–5
Cold Spring Harbor 113, 116, 119
Collège de France 80
Columbia University, New York 146–7, 148, 150, 151–2, 155, 157
Comptes Rendus 116
computing 52, 62, 132, 135, 137, 138, 150, 176, 181–2, 184–92
Comrie, Leslie John 'L.J.' 88, 89–90
Control Data Corporation (CDC) 192
Copenhagen University 55, 56, 58, 59, 65
Cornell University 104–9, 117, 198, 199–200
Creighton, Harriet 107–8
Crick, Francis 168, 169, 170–2, 174
Croix de Guerre 83
Crowfoot Hodgkin, Dorothy 121–140, **129**, 170
  career 128–40
  children 133–5, 136
  education 122–3, 124–7, 132
  illness 121, 131, 133–4, 136, 140
  marriage 132–3
  research 122–3, 126–40
Crowfoot, John and Mary 'Molly' 122, 123, 124–5, 126, 127, 138
crystallography 126–8, 134–40, 143, 163, 165–71, 173–6
Curie, Eugéne 67–8
Curie, Eve 68, 69, 70, 74
Curie, Marie 20, 67–72, 73, 74, 80, 84, 138, 145, 164, 197
Curie, Pierre 67–8, 69, 73, 84

Dainton, Fred 164
Dalton, John 178
Danish Geodetic Institute 60–2, 64, 65
Danish Geodetic Service 60
De Saussure, Horace 4–5
De Vaucouleurs, Gérard 202–3
Debierne, André 73
Demerec, Milislav 113
Department of Agriculture, Washington DC 196
Department of Applied Economics (DAE), UK 189–91
DNA (desoxyribose 'deoxyribose' nucleic acid) 106, 108, 114, 116–18, 156, 159, 168–73, 176

Earth, inner core of 62–4
economics 184, 188–92
Eddington, Arthur 87, 88–9
Edinburgh New Philosophical Journal 11–12
Eighth International Congress of Crystallography 139
Einstein, Albert 17, 21, 23, 29, 35, 39, 44, 46, 48, 54, 69, 70, 87, 144, 148, 150, 154
Electronic Delay Storage Automatic Calculator (EDSAC) 184, 188
Eliot, T.S. 139
Emerson, Rollins 109–110
Emil Fischer Medal 23–4
Enrico Fermi Award 38
Erlangan, Germany 39–40
Essex, University of 190
Ettore Majorana Centre for Scientific Culture, Italy 158
Eunice Newton Foote Medal for Earth-Life Science, The 15
Exner, Franz 18, 19

feminism 2, 55, 159
Ferdinand, Archduke Franz 70
Fermi, Enrico 30, 33
Ferry, Georgina 125, 126, 137
Festival of Britain 181
Feynman, Joan 94
Feynman, Richard 94, 110, 119, 156, 198, 200
Finch, John 174
First World War 23, 43, 45, 59, 70–1, 86, 104, 122, 160

Foote, Elisha 2–3
Ford, Gerald 157
Ford, Kent 207–12
Fort Halsted, UK 182
Fourier, Joseph 4–5, 6
Franklin née Waley, Muriel 160
Franklin, Ellis 160–2
Franklin, Rosalind 'Ros' 'Roslind' 159–77, **167**
  career 165–76
  death 159, 176
  education 160–5, 166
  illness 175–6
  research 165–76
Freedman, Wendy 213–14
Friends Voluntary Service for Unemployed Miners 133
Frisch, Otto 25, 29, 34–5, 36, 38
Fry, Margery 125
Furness, Caroline 198

Gamow, George 34, 202
Gaposchkin, Sergei 98–9
Garrod, Dorothy 162
Gay-Lussac, Joseph 13
Geiger counter 147
Gell-Mann, Murray 156
genetics 105–6, 107–9, 113, 114–19
Genetics Society of America 112, 114
Georgetown University 201–7
German Astronomical Society (Astronomische Gesellschaft) 98
Germany 23–4, 28–31, 39–40, 45–6, 53–4, 81, 98, 109, 127, 141, 145, 161–3
Gibbs, Josiah Willard 119
Girton College, Cambridge 36, 56, 161, 183
Gonella, Nat 178
Gordan, Paul 40, 43
Gosling, Raymond 168–71
Göttingen, Prussia 40, 42–4
Gray, Frances 86
greenhouse effect, terrestrial 3–9
Griffiths, Ada 160
Groes, Nils 65
Gu Jing-Wei 143
Guggenheim Fellowship 109
Gutenberg, Beno 60, 64

H.D. Cooke Elementary School 196
Hahn, Otto 20–4, **27**, 29–31, 33–4, 35, 36–8, 81
Haldane, J.B.S. 162
Hanford Site, Washington State 147
Hartmann, Johannes 45, 46
Harvard College Observatory 89–91, 96, 100
Harvard University 89–100
Harwell, UK 182
Haverford College 201
Hawkins, Coleman 178
Heisenberg, Werner 25, 29, 148
Henry Norris Russell Prize 101
Henry, Joseph 10, 12
Hermann, Grete 52
Herschel, William 6
Heyden, Father Francis 201, 202
hidden variables theory 52
Hilbert, David 40, 43, 44–6, 48
Hitler, Adolf 28
Hodgkin, Thomas 132, 134, 136, 140
Hoffmann, Adolph 46
Hollerith machine 135
Holmes, Kenneth 174
Holst, Gustav 86
Hong Kong 156
Hood, Dorothy 128
Hoyle, Fred 186
Hubble, Edwin 205
Illinois, Urbana-Champaign, University of 202
Imperial Chemical Industries (ICI) 131, 168
Independent Social Democratic Party (USDP) 46
Institut de France 69
Institute for Advanced Study, Princeton 152
Institute of Theoretical Astronomy 186
insulin 132, 138, 139–40
International Astronomical Union (IAU) 204
International Computers Limited (ICL) 190
International Day of Women and Girls in Science 157
International Union of Crystallography 140

Jacob, François 116
Jacobsen, Lif Lund vii, 55, 58
Joliot-Curie, Hélène 74, 76
Joliot-Curie, Irène 30, 31, 33, 67–84, **75**, 138
  career 71–4, 76, 84
  children 74, 76
  education 68, 70–2
  marriage 72, 74
  nursing 70–1
  research 76–82
Joliot-Curie, Jean Frédéric 'Fred' 30, 72–84
Joliot-Curie, Pierre 76
Joseph, A.F. 123

Kaiser Wilhelm Institute (KWI) 21–3, 109
Kaldor, Nicholas 190
Keller, Evelyn Fox 105, 108
Keyser's bank 160
Kimber Genetics Award 117
King, Mervyn 189

King's College, Cambridge 163
King's College, London 168–9
Kitt Peak National Observatory 205
Klein, Felix 40, 43, 46
Klein, Oskar 37
Klug, Aaron 174, 176

Laboratoire Central des Services Chimiques de l'Etat (now the Institut National de Recherche Chimique Appliquée), Paris 165, 170
Lamont Geological Observatory, USA 64
Langevin, Paul 68, 69–70, 71, 73
Lawrence, Ernest O. 144, 145, 146
Le Quotidien 72
League of Nations 123
Lee, Tsung-Dao 151–3, 155
Lefschetz, Solomon 53
Lehmann discontinuity 64
Lehmann, Alfred 55, 58
Lehmann, Inge xiv, 55–65, **61**, 67
  career 58–62
  education 55–60
  research 60–5
Lick Observatory, California 97, 98
London, University of 191
Lowell Observatory 214
Lowell, Abbot Lawrence 97, 100
Lyman, Theodore 95

Maddox, Brenda 160
Magnus, Gustav 9
Makemson, Maud 198, 199, 202
Manchester University 182
Manhattan Project 145, 146–7
Mao Zedong 147
Mary Ewart Scholarship for Natural Sciences 87
Massachusetts Amherst, University of 213
mathematics 39–54, 56–7, 59, 86, 156, 180–1, 182, 197
Mathieu, Marcel 165
McCarthy, Martin 205
McClintock, Barbara 103–19, **111**
  career 106–15, 117, 118–19
  education 104–6
  research 106–19
McClintock, Thomas and Sara 103–4
Medical Research Council 135, 171, 176
Meitner, Lise 17–38, **27**, 39, 78–9, 81, 141, 145
  Anschluss 31–2
  career 19–25, 32–3, 35–6, 38
  education 17–19
  medical work 23, 28
  research 19, 22, 25–8, 30, 32–6
  retirement 38
  writing 21
Meitner, Philipp and Hedwig 18
Mendeleev, Dmitri 30
Menzel, Donald 100
Mering, Jacques 166
Meyer, Leopold 58
Meyer, Stefan 19
Michelson–Morley experiment 155
Michigan, University of 143–4
Military Medal 71
Minkowski, Hermann 40
Minnesota, University of 213
Missouri, University of, Columbia 65, 110–12
Mitchell, Maria 198
Monod, Jacques 116
Morgan, Thomas Hunt 110
Moszkowski, Steven 156
Mount Wilson Observatory, California 98, 205
Murray, Betty 127–8
Mutch, Eric 181

National Academy of Sciences, USA 112, 114, 117, 155, 203
National Bureau of Standards (now the National Institute of Standards and Technology), Washington DC 153, 202
National Institutes of Health (NIH), USA 175, 175
National Medal of Science 117
National Research Council (NRC) 108
National University (now Nanjing University), China 143
Nature 36, 131, 132, 171–2, 206
Naturwissenschaften 33
Nazi Party 29–9, 53, 83, 98, 130, 161–2, 163, 164
New York Times, The 54, 100
Newnham College, Cambridge 56–7, 87, 130, 161–2, 184
Newton Foote, Eunice 1–15, **7**, 113
  education 2
  inventions 14
  marriage 2–3
  publication 10–12, 13–14, 15
  research 3–6, 12
Newton, Isaac 2, 204
New-York Daily Tribune 1
Niels Bohr Institute for Theoretical Physics 24
Nobel Committee 17, 36–8

Index

Nobel Prize 29, 32, 33, 36–8, 67, 68, 69–70, 71, 76, 77–80, 110, 118, 121, 124, 135, 137, 138, 140, 141–2, 144, 148, 151, 155, 157, 172, 176, 177, 189, 191
Noether, Emmy 39–54, **47**
  career 41, 43–8, 51, 52–4
  death 54
  education 40–1
  habilitation 41, 43–6
  Noether's theorem 49–50
  research 42–3, 48
Noether, Max and Ida 39, 43
Nørlund, Niels 59–61
nuclear fission 17, 34–7, 82–3
nuclear physics 143–8, 150–1, 152–4, 156, 181
nuclear reactor 147
nuclear testing 64–5
Nuffield Foundation 137

Ohio State University 213
Oldham, Richard 63
Ontario County, New York State 2
Oppenheimer, J. Robert 145
Order of Merit 139
Oxford University Labour Party 133
Oxford, University of 53, 121, 124–8, 130–2, 163, 182

Palestine 160
Palomar Observatory 205–7
particle accelerators 32, 144, 150, 154
Paul Junior High School 197
Pauli, Wolfgang 36
Pauling, Linus 137, 163
Payne-Gaposchkin, Cecilia 85–101, **93**, 109
  career 96–101
  education 85–9, 94–6
  marriage 98–9
  research 90–6, 99–100
penicillin 113, 134–5
periodic table 30
Perrin, Jean 68, 82
Perutz, Max 37, 140
Philipp, Kurt 26
Philosophical Magazine 14
Physical Review 151
Physics Today 213
Physics World 37
Pilot ACE computer 137
Planck, Max 19–20, 22, 29
Poland 163
Portsmouth technical college 178
Powell, Herbert 'Tiny' 127, 136
Princeton University 146, 156
Princeton University Observatory 95
Proceedings of the American Association for the Advancement of Science 13, 15
Proceedings of the National Academy of Sciences 108, 116
Pulkova (now Pulkovo), Russia 98
Pusey, Norah 123

quantum physics 25, 50, 51, 52, 92, 148–54, 163

Radcliffe College 95
Radiation Lab, Berkeley 144–6
radioactivity 67–70, 73, 76–83, 144–5, 146–7, 152–3
Radium Institute, Paris 70, 71–2
Ramsay, William 20
Randall, John 168, 170, 171
RCA (Radio Corporation of America) 146
relativity theory 17, 40, 44, 46, 87, 144
Religious Society of Friends (the Quakers) 90
Rhoades, Marcus 107, 110, 112
Richter, Charles 64
ring theory 50–1, 52
RNA (ribosenucleic acid) 173
Robinson, Robert 131
Rockefeller Foundation 53–4, 110–11, 134, 136, 137, 174
Rosenbeck, Bente xiii–xiv, 103
Rossiter, Margaret xiv
Routledge & Kegan Paul 160
Royal Greenwich Observatory (RGO) 206
Royal Institute of Technology, Sweden 38
Royal Institution, London 6, 123, 128, 165, 175
Royal Medal of the Royal Society 138
Royal Society, London 65, 133, 135, 138, 140, 174
Royal Swedish Academy of Sciences 32, 36–8, 69
Rubens, Heinrich 20
Rubins, Bob 198–9, 214
Russell, Henry Norris 95–6
Rutherford, Ernest 20, 21, 69, 77, 87, 145
Ryle, Martin 206

Saha, Meghnad 91–2
Samuel, Herbert 160, 164
Sandage, Allan 205–6
Savitch, Paul 80–1
Schrödinger, Erwin 25, 29, 148, 181, 184
Schwarzschild, Karl 40
Science and Technology Policy, Office of (US) 157
Science for Peace prize 158
Scientific American 11, 15, 174

Scientific and Industrial Research, Department of (UK) 165
Second World War 36, 64, 82–3, 134, 145, 161–5, 178–81, 197
Segrè, Emilio 144
Seismological Laboratory of the California Institute of Technology 64
seismology 60–5
Seneca Falls Convention 1848 3
Shaknov, Irving 150
Shapley, Harlow 89–90, 92, 94–8, 99, 100, 197
Shaw, William 200
Siegbahn, Manne 32
Signals Research and Development Establishment (SRDE) 181
Signer, Rudolf 168–9
Sime, Ruth 32
Sixth International Congress of Genetics 108
Slater, Lucy 176–193, **187**, 195
  career 179–91
  education 178, 179, 180, 181, 183–4
  injury 180, 188
  music 178–9
  research 181–91
Smith College, Massachusetts 146
Solvay Congress *or* Conference 30, 78
Solvay, Ernest 69
Somerville College, Oxford 124, 128–133, 135
Sorbonne, Paris 70–1
Southampton 178–80
Southampton University College 180, 181
Stadler, Lewis 110–12
Stahr, Martha 199–200
Stanford University 113, 213
Stanton, Elizabeth Candy 3
Steffensen, Johan 60
Stevens, Katie 122
Stone, Richard 189, 191
Strassmann, Fritz 29, 31, 33, 35, 37, 38
suffragettes 70, 178, 182
Switzerland 151

Taiwan 151, 156
temperance 2
thermodynamic laws 18
Thomson, J.J. 162
Turing, Alan 137, 182
Tyndall, John 6–9, 14

US Army Air Force 186
US Army Signal Corps 197
US Geological Survey 213
US Naval Research Laboratory, Washington DC 198
USDP *see* Independent Social Democratic Party

Van der Waerden, B.L. 52
Vassar College 197, 202
Vera Rubin 195–214, **209**
  career 201–5, 213–14
  children 200–1, 202, 203, 213, 214
  education 196–208
  marriage 198–9
  research 198, 199–200, 201–13
Vienna 17–18
Vienna, University of 18–19
virus research 134, 173–6
vitamin B12 136–8
Von Neumann, John 52, 119

Watson, James 169, 170–2, 173, 174
Weill, Adrienne 164, 165
Wellcome Laboratory 123
Wells, David 10
Wheeler, John 150
Whitney, Mary 198
Wilkins, Maurice 168–9, 171–2, 174
Willard, Emma 2
Wilson, Harold 190
Winchester High School 180–1
Wolf Prize 157
Wolfson, Lord 138
women's rights 2–3, 15, 17, 18, 19–20, 40–1, 43–5, 46, 53, 54, 55–9, 70, 72, 78, 82, 89, 94, 105, 109, 114, 125–6, 130, 139, 144–6, 156–8, 163, 169, 174, 177–9, 185, 188, 189, 191, 195–6, 204, 205–6, 212–13
Woods Hole 173
Workers' Educational Association 134
Working Men's College 160
Worldwide Standardized Seismographic Network (WSSN) 64–5
Wu Zhong-Yi and Fan Fu-Hua 142–3
Wu, Chien-Shiung 141–58, **149**
  career 144–8, 151, 155–8
  children 148
  education 143–5
  marriage 146
  research 144–56

X-rays 18, 70–1, 124, 126, 127, 130–2, 134–6, 143, 163, 166–72, 174–6

Yang, Chen-Ning 151–3, 155
Yerkes Observatory 99
Yuan, Jada 142, 145, 147
Yuan, Luke Chia-Liu 144, 146

Zeilinger, Anton 151
Zwicky, Fritz 210